D0425439

Advance Praise for

ONE SIZE FITS ONE

"If there is one book I recommend you read from cover to cover, it's this one. *One Size Fits One* introduces you to today's customer—a customer whose needs can only be met by a loyal and committed cadre of enthusiastic employees."

Shannon Clyne, former Chief Investment Officer
Bank of America

"At last a book about business that recognizes the most important investment to be made is in human capital. Gary not only captures the key to success in business, but also the key to success in life."

Eunice J. Azzani, Vice President
Korn/Ferry International

"Unfaithful to its title, this book is a perfect fit for anyone who wants to grow his or her business."

David Samec, President
Samec Associates

ONE SIZE FITS ONE

Building Relationships One Customer and One Employee at a Time

Gary Heil
Tom Parker
Deborah C. Stephens

Van Nostrand Reinhold

I(T)P® A Division of International Thomson Publishing Inc.

New York • Albany • Bonn • Boston • Detroit • London • Madrid • Melbourne
Mexico City • Paris • San Francisco • Singapore • Tokyo • Toronto

The ideas presented in this book are generic and strategic. Their specific application to a particular company must be the responsibility of the management of that company, based on management's understanding of their company's procedures, culture, resources, and competitive situation.

Printed in the United States of America

http://www.vnr.com Visit us on the Web!

For more information contact:

Van Nostrand Reinhold
115 Fifth Avenue
New York, NY 10003
USA

Chapman & Hall GmbH
Pappalallee 3
69469 Weinham
Germany

Chapman & Hall
2-6 Boundary Row
London SEI 8HN
United Kingdom

International Thomson Publishing Asia
60 Albert Street #15-01
Albert Complex
Singapore 189969

Thomas Nelson Australia
102 Dodds Street
South Melbourne 3205
Victoria, Australia

International Thomson Publishing Japan
Hirakawa-cho Kyowa Building, 3F
2-2-1 Hirakawa-cho, Chiyoda-ku
Tokyo 102 Japan

Nelson Canada
1120 Birchmount Road
Scarborough, Ontario
M1K 5G4, Canada

International Thomson Editores
Seneca, 53
Colonia Polanco
11560 Mexico D.F. Mexico

2 3 4 5 6 7 8 9 10 QUEBP 01 00 99 98 97

Library of Congress Cataloging-in-Publication Data available upon request.

ISBN 0-442-02063-5

Text Design: Keithley Associates, Inc.

Production: Jo-Ann Campbell • mle design • 562 Milford Point Rd., Milford, CT 06460

Contents

Part Three

Building Relationships One Employee at a Time • *181*

FOREWORD

A Succession of Moments

There are two great motivators in life. One is fear. The other is love. You can manage by fear, but if you do you will ensure that people don't perform up to their real capabilities. But if you manage people by love—that is if you show them respect and trust—they perform up to their real capabilities. They dare to take risks. They can even make mistakes. Nothing can hurt.

—Jan Carlzon

Jan Carlzon is former Chief Executive Officer, President and Chairman of Scandinavian Airlines (SAS). Having led the rebirth of the company on a scale of Iaccoca's resurrection of Chrysler, Carlzon literally turned the organization's structure upside down and within twelve months the company went from near bankruptcy to profitability. The World Executive Digest *named Carlzon one of the ten best leaders in the world.*

In 1987 Carlzon's book, Moments of Truth, *became an international best seller. In the book he described the fleeting contact between a business and its customers as a critical moment of truth. The concept has since become a service classic and, today, remains at the foundation of many service training programs and books for numerous industries throughout the world.*

My inspiration for writing *Moments of Truth* sprang from a simple observation made during my tenure at Scandinavian Airlines: that the essence of any organization is communicated in the single fleeting moment when someone in that organization connects with a customer. This communication takes place, for example, when a flight attendant brings you your coffee, a receptionist bids you good morning, or a salesperson calls to tell you your order has arrived. Taken together, these small, seemingly inconsequential moments define who you are and establish your worth in the marketplace. The greater the number of positive moments you have, the greater the value of your enterprise. The fewer you have, the lower its value.

Today, with the wide array of choices open to customers, these brief encounters are more critical than ever—not only for what they say about you as a company but for how they affect your ability to build the long-term relationships with your customers that, I believe, are imperative if your organization is to remain profitable. To create such continuity requires that every moment with every single customer builds to another until a succession of these moments leads to a powerful relationship that, from your customer's perspective, is both compelling and irreplaceable.

Most of these moments will arise naturally in the course of your day-to-day business—a telephone call confirming an appointment, an e-mail inquiring about an overdue invoice, a request for a change in an order. Equally important, however—and often overlooked—are those moments created proactively by a business that demonstrate its interest and concern for its customers. The CEO of Ericsson Telecommunications, for example, spends fully half his time with existing customers to ensure that they remain pleased with the company's products and its follow-on service.

In most medium- to large-sized businesses, of course, these moments rarely involve the CEO. Instead, the overwhelming majority of these critical points of contact are made by front-line employees. These are the men and women who provide the foundation for the relationships a company builds and, as such, whose actions determine whether the company lives or dies. And it is these individuals—along with its customers—that are any company's greatest asset. Equipment and real estate may be bought and sold, but the company with excellent employees enjoys the most priceless, hard-to-find, and sought-after asset.

For these assets to deliver full value and appreciate over time, however, they must first share a vision of the company's future. Consider the story of the two stonecutters: The first stonecutter was miserable in his work and, when asked why, said, "All day long, I just chip away at these lousy rocks." The second stonecutter performed exactly the same task, yet seemed excited by what he was doing. When asked about his excitement, the second man said, "I'm part of a wonderful team that is crafting raw materials to construct a magnificent cathedral." In other words, without a vision to provide meaning to their actions, employees will be hard-pressed to create a positive moment. With a vision, they can't help but be excited by what they do—and they communicate that excitement to those with whom they come into contact.

Likewise, employees haven't a chance to live up to their full potential without information and the authority to use it in their work. On the other hand, *with* information and the blessing of management to do with it what each employee feels is best, employees are usually effective and responsible. If the moments your customers experience are to be positive ones, then every employee who plays a role in those moments must have the power to control their outcome.

By way of illustration, not long ago I asked the manager of the Thai Hotel in Bangkok how his leadership style helped facilitate the excellent service I had encountered on my stay. Given my belief in employee empowerment, I was astonished when he told me that he achieved this result by limiting front-line authority "How so?" I asked. "Well," he said, "I've taken away virtually everyone's authority to say no to any customer's request. The only authority people have is to say yes." Would that most organizations were so authoritarian!

In any event, I'm absolutely convinced that you cannot manage an organization by withholding power or through fear for, if you do, you will ensure that people won't perform up to their real capability. If, however, you manage by respect and trust, people will flower and grow—as will your business—and relationships will thrive.

I find it gratifying and entirely appropriate, therefore, that the focus in *One Size Fits One* falls on both customers *and* employees, and that authors Gary Heil, Tom Parker, and Deborah Stephens hold that building long-term relationships with good customers is the true and logical path to the successful future of the organization. In the One Size Fits One world so skillfully and thoroughly evoked on these pages, singular "moments of truth" are strung together into a series of connected events that reinforce and deepen individual relationships to the ultimate benefit of both the customer and the organization.

To be sure, a successful venture requires more than these positive moments alone. For example, to succeed in the millennium will also demand that companies cut waste and work at peak efficiency. But, even if waste were eliminated completely and a company ran as smoothly as a Swiss watch, success would not be guaranteed. On the contrary, a merely "efficient" organization is like a body with a head but no heart. Strong relationships, as the authors point out, can exist only if both heart and mind are taken into account. You simply cannot have one without the other.

Also, I find it refreshing and enlightening that the authors have chosen to tell their cautionary yet optimistic tale using the voice of an informed customer. Surely, in a world where keeping your existing customers is critical, it is absolutely imperative that we know what is going on in their minds. And who better to tell us than those customers themselves?

Finally, it would be unthinkable to write a book about relationships and service without more than a passing mention of the contribution of employees to this challenging effort. Here, again, the authors have repeatedly placed the emphasis precisely where it must be—on the men and women in the organization who form those relationships and tend them on a daily basis. It is virtually impossible to overstate just how important each and every employee is to the success of our organizations—particularly these days when we are calling on these individuals to shoulder increasingly complex responsibilities. For our organization to reach its goals, each employee must think and act as if he or she owned the company...and must have the power and confidence to do so.

Looking at American companies, I am almost always impressed and heartened by what I see, particularly by their abilities as service providers. The challenges of the future, as outlined in this book, however, will test even the best of them as never before. But I am confident that with books like *One Size Fits One* to serve as both guide and beacon, the pioneer spirit and entrepreneurial fervor that have always been one of the hallmarks of American business will enable companies worldwide to thrive in the years to come and to lead the way into the next century.

Jan Carlzon
Stockholm, Sweden
September, 1996

ACKNOWLEDGMENTS

GARY HEIL...

Thank you Carol, Ryan, and Michaela for your encouragement and love. Your support gave this project meaning.

My partners, Deborah Stephens and Tom Parker, were great friends and great teachers. Our discussion continually challenged me to re-think my thinking. I'm very lucky to have been part of the team.

To Don Peppers and Martha Rogers, thank you for leading the demassification revolution. As they're quick to point out, the one to one future is now the present.

This book has Rick Tate's fingerprints throughout. It was his idea to use the customer's voice when describing customer relationships. He felt that both the writers and the readers would stand a little more firmly in the customer's shoes that way. As *usual* he was right. Rick has been my favorite teacher for decades. I could never thank him enough.

I would not be in a position to write this book if it were not for the friendship and support of Ken Blanchard. Ken's example, teaching, and good humor provided direction when it was needed most. Thanks again, Ken.

Honesty, integrity, competence, empathy—are the words that come to mind when I think of the people at the Washington Speaker's Bureau. Their support changed my life. Their example has made me a better person.

Thank you, Georgene Savickas, for your humor and encouragement. But most of all thank you for teaching me how to face adversity with strength and faith.

Thanks to John Boyd, Chris Bates, and the team at VNR for their one-size-fits-one support.

I've been fortunate to have been asked to work with many leaders who *"live"* what we *"talk"* about. To each of them, thanks for the support and the education.

TOM PARKER...

With time short, space limited and so many people to thank for their support, I am left no alternative but to offer only one-size-fits-all acknowledgments—with the expectation, of course, that these individuals (you know who you are) will remain loyal:

For my students, who are my best teachers;
For my clients, who are my soundest advisors;
For my old friends, who are my most tolerant audience;
And for my family, who are my delight and the source of my strength.

DEBORAH COLLINS STEPHENS...

To my children, Aaron and Lily, thank you for being my source of inspiration, unconditional love, and enduring optimism.

To Gary Heil, your deep regard for the human spirit, your unrelenting belief in people and your contagious optimism have been a joy to witness firsthand. Your words, whether written or spoken, touch us all in that part of our souls where our dreams lie. Thank you for giving me the opportunity to work with you and to learn from the tremendous contributions you make to organizations and people around the world.

To Jackie Speier and Linda Allan, for teaching me the true meaning of service. Your ability to lead with courage, your wise counsel, and your never-ending, always comforting, support and encouragement has enriched my life and made me a better person.

To John Boyd, Chris Bates, Jacqueline Jeng, Angela Burt-Murray and Marc Sperber of Van Nostrand Reinhold, thank you for your support, encouragement, and belief in this project.

To Steve Sierra, whose life of service to others lives on in each who knew him.

PART ONE

Welcome to a World Where One Size Fits One

Something's happenin' here...what it is ain't exactly clear...

—Buffalo Springfield

CONSIDER THE FOLLOWING:

- A billion dollar paper manufacturer in Wisconsin works closely with a small stationery store halfway across the country to better ensure that the company's products will sell at the retail level.
- A clerk at a notions store in Mountain View, California, calls a supplier long-distance to check on the availability of a single 75 cent button for a first-time customer.
- An electronics equipment manufacturer in South Carolina offers a brand-new but up-and-coming company its best discount for a *single* item—incurring a significant loss on the transaction—with the hope of winning the client's business over the long term.

- An irate customer in Berkeley, California, places a $10,000 ad in the *Wall Street Journal* to protest what he considers shoddy treatment by a large corporation—and ultimately receives 6,000 responses from other dissatisfied customers over his toll-free telephone number.

Love it, hate it, fear it, or wish it would just disappear, we are entering an era where one size no longer fits all—or even a few. We are entering an era where One Size Fits One. It's a highly personalized, customercentric, customer-driven time. A long-term relationship with each profitable customer is the primary objective, and customers collaborate with you in ways that can build your business…or, if they're sufficiently dissatisfied, frustrated, or angry, work against you to undermine it.

RELATIONSHIPS ARE THE CURRENCY OF THE FUTURE

It's also a world where success isn't measured only by how *much* you sell or the percentage of the market you own but by how *deeply committed* your best customers are. A world where the emphasis is less on building short-term satisfaction than on instilling long-term loyalty. And a world where relationships are the currency and the committed employees who maintain and nurture them are the primary source of competitive advantage.

More than (yet another) service tweak or a new front end on the way most of us are already doing business, this new world, with its focus on individual treatment and long-term commitment, is a fundamentally different approach to the way we serve customers and the way we go about building mutually beneficial relationships with them…as well as with our co-workers and our strategic partners. Indeed, in this rapidly emerging business environment, what we used to think of as "service" turns out to be merely the tip of the iceberg.

Meanwhile, lurking beneath the surface are the more challenging issues involved in identifying the customers with whom we want to do business, creating a work atmosphere in which relationships can thrive, selecting and educating employees who genuinely care about each and every customer, and then ensuring that they have the responsibility and the authority to keep those customers in the fold. In a One Size Fits One world, we move a serious step beyond customer service. Instead, we must now turn our energies toward creating an even more complex relationship culture.

At its most profound, the arrival of a One Size Fits One world signals the end of the era of mass production at the same time that it serves as the bellwether of a new era of customized service, deeply held mutual commitments, and specialized treatment for customers and employees alike. *Out* is the mass production mentality that has prevailed in this country ever since the first Model T rolled off the assembly line and people like Adam Smith and Frederick Taylor helped redefine the worker as a cog in the complex corporate machine. Out too is mass production itself, along with mass marketing, mass merchandising, and virtually "mass" anything...just as Alvin Toffler predicted when he coined the term *demassification* nearly 20 years ago.

In, and in their place, are products and services—from pants to pizzas and bank cards to insurance policies—tailored to each individual customer's need. Personalized attention and support and, above all, the ability of each and every customer to derive something unique from everyone he or she does business with are the prerequisites of success. In a world where One Size Fits One, no one will have to settle for the ordinary—and no business will be able to survive by providing it.

Today's business environment marks the end of nearly a century in which customers had to *take* essentially what companies had to *give,* and ushers in a time when good customers will be able to tell businesses what to make, when to make it, and how to deliver it. Where once manufacturers were able to *push* their products and services to a relatively acquiescent marketplace, in the future, customers, as the only true and legitimate arbiters of value, will create the *pull* that will determine the direction and focus of even the largest corporation. The way value will be conceived, produced, and delivered will be turned on its ear, and a whole new set of rules—as well as rulers—will emerge. Neither the way we buy nor the way we do business will ever be the same.

A POWER SHIFT

If you don't quite see this megatrend in the making, or if the change in the marketplace today doesn't seem as extreme or radical to you as we've just outlined, then, before you dismiss or ignore it, we urge you to take a closer look at the savvy companies that are focusing their energies on building relationships and at the equally savvy customers who are consolidating their newfound power to get more of what they want.

Like an offshore hurricane, it is gradually building strength. And, once it finally hits the mainland with its full force, it will have an impact and reverberations the likes of which the marketplace hasn't seen since the Industrial Revolution. So, if you have yet to feel the winds of One Size Fits One blowing through your corridors, don't be fooled—it's only the calm before the storm. It will be coming soon to your industry...and to your community.

What makes this force so powerful and inevitable and, for customers, so beguiling? Any number of marketplace forces and technical developments, not the least of which are information and manufacturing technologies that enable the cost-effective customization of products and services, along with worldwide increases in competition for customer dollars. Better processes, stronger strategic partnerships, more responsive employees with greater responsibility, and the acknowledgment by most forward-looking companies of the need to change in order to grow are all contributing to the inevitability.

Weighing in most heavily, however, in the imminent arrival are two key factors at once discrete and related. The first of these is the realization by many companies—bolstered by recent research—that *having loyal customers is critical for long-term survival.* With an ever-diminishing supply of so-called "good customers"—those who will buy more over time, pay a premium for exceptional service, send their friends, and play by a reasonable, mutually arrived-at set of rules—more and more companies are coming to understand that the best way to grow their business is by deepening the relationships they already have rather than constantly looking for new ones.

Not only have we finally learned that, in an era of tough competition and tight margins, losing old customers as quickly as we find new ones is too expensive, we have also realized that the all too familiar cycle of customers in/customers out exacts other, more subtle costs. Without retained customers to pay the bills and keep profits and company spirits up, for example, good employees are hard to find and keep. Aggressive pricing policies are difficult to implement, and long-term shareholder interest is almost impossible to come by. Also, as good customers defect one by one and commit to other companies, the gene pool of remaining customers becomes depleted, leaving those companies that have been abandoned with the dregs. Or, alternatively, no customers at all.

To say that this scenario has forward-looking companies across the country nervous would be an understatement. In-the-know businesses of all sizes and types are scrambling to take the next step to build a loyal customer base. In their advertising, in their correspondence, in their annual reports, and in their day-to-day dealings with us as customers, they have as much as told us what they want: our business today *and* tomorrow. And the way most of them have chosen to go about getting that business is not simply by cutting prices or improving product quality—although cost and reliability will always be factors in the value equation. Rather, they're shoring up their relationships with their best customers, *one customer at a time,* and seeing to it that those customers will be around for the long term. By focusing first and foremost on building these relationships, these companies are, in essence, changing to meet the basic requirements of a One Size Fits One world, and hastening its arrival in the process.

Meanwhile, customers are developing their own agendas. Rarely having been asked to be loyal to a company before—and suddenly inundated by entreaties for their commitment—customers aren't exactly flocking to give companies what they're seeking. Instead, they're considering their options, reluctant to give their loyalty cheaply or without a significant commitment in return.

This set of circumstances, in turn, is leading to a gradual shift in power from those who create products and services to those who consume them. This *power shift* in its various manifestations is fueling the revolution that will force companies to renew themselves quickly or become extinct.

As a result, from the customer's perspective, things have never looked more promising. Nor have customers ever enjoyed as much power. Indeed, as businesses put on a full-court press for customer loyalty, as competition grows and improves, and as customers become better educated and more sophisticated about their choices, for the first time *ever*, the customer is finding himself or herself in the catbird seat.

To be sure, this power shift is no overnight phenomenon. On the contrary, it's one of those subtle, gradual movements that's hard to discern from one day to the next. Also, it is still in its infancy. Nevertheless, examples of this power shift are everywhere in evidence, with none more direct and graphic than customers' growing willingness to vote with their wallets. Frustrated by years of having to take what companies chose to dish out, customers are making their feelings known by taking their business elsewhere. They're frustrated and they're taking action.

COLLECTIVE CONSUMER CONSCIOUSNESS

Better informed about what's happening in the business world and more attuned to the potential influence in the marketplace of radio, TV, and print, as well as the newly emerging voice of the Internet, customers today have also developed a broad-scale awareness of their power which, 10 years ago, was recognized by only a few media-wise activists. As a force, this awareness, or *collective consumer consciousness,* tells customers in no uncertain terms that they can have it their way and that, if they're not getting it, they should hold out until they do. That it's possible to take on a business or company of any size and, rather than be ignored, come out a winner. That it's possible to formulate, declare, and act on their own "personal policies" to combat "company policies" that they feel are ridiculous or unfair. And, finally, that it's possible to join in the ever-growing crowd of customers who refuse to have their requests or opinions standardized, patronized, or minimized.

The power of this widely held awareness to affect how business will be done should not be underestimated, particularly when coupled with customers' long-standing frustration with most businesses and their collective and individual reluctance to give their loyalty just at a time when companies need it most. In a one-size-fits-all world, there seems to be no resolution to this disjunction, no path that can straddle what today's more aware and better-educated customer wants and what tomorrow's profit-seeking businesses have to have.

With customers and companies working together, however, both needs can be met—with the individual relationship that results between company and customer serving as the basis on which a mutually creative value can develop.

PUTTING A FACE ON THE FACELESS CUSTOMER

Meanwhile, other companies like Disney and Seattle-based high-end retailer Nordstrom, as well as a number of newer companies, such as Ritz-Carlton and Southwest Airlines, already had cultures based on building strong customer relationships and found themselves in the fortuitous position of being in the right place at the right time. There is, in any case, little argument that, over the last decade, product quality has skyrocketed, products have become more reliable, and, in many cases, their costs have gone down. That there has been a marked improvement in service is undeniable.

What essentially happened to create these new levels of quality and service was that, in the mid- and late-1980s, many companies took a closer look at what their customers perceived as *value*—the equation that balances product or service quality and reliability, delivery time, overall responsiveness, and, of course, price—and revamped their operations to better deliver that value. In response to the customer demand for change, and coupled with the pressing, profit-driven need to develop a loyal customer base, companies expended huge amounts of energy to improve, and in some cases even customize, production to appeal to an ever-larger number of customers. The name of the game was market share, and companies did whatever they had to do to win it: faster, smaller, better, and cheaper computers; better-built cars that were less expensive to maintain; insurance claims paid more promptly.

While these efforts frequently resulted in the sought-for improvement in products and services, much to the surprise of those providing them, these improvements often *didn't* result in significant increases in customer loyalty. In fact, as far as generating loyalty went, many companies found that they had, inexplicably, hit the wall.

In retrospect, the failure to build loyalty shouldn't be surprising, particularly because, at the same time that we were improving our products and processes, customers were becoming aware that they could demand individual treatment and enjoy a considerably more beneficial relationship with businesses than they had been able to in the past.

To say it straight, most of us involved in service businesses underestimated the challenge. We failed to recognize how we'd have to undergo fundamental change in order to build loyalty among customers. We often paid less attention to relationships per se and more attention to the rational aspects of the value equation. Most important, we had not sufficiently factored into our practices the most critical element in virtually all relationships, business relationships included—the human element, *emotion.*

BUSINESS IS A HUMAN ENDEAVOR

We simply tended to overlook the essential facts that, at its heart, business is a human endeavor where individuals meet, talk, work, and otherwise try to help and benefit one another and that emotions were and are *at least* as much the currency of exchange, satisfaction, and loyalty as dollars. Messy, elusive, irrational, and difficult to quantify, the emotional component of the value equation had been largely ignored—and often for these very reasons.

Sure, we were making things faster, better, smaller, and cheaper. But, for all that, we often didn't make the business interaction sufficiently satisfying on a *human* level to keep customers coming back and recommending us to their friends. Why? Because, in our haste to provide for *all* customers, we overlooked the far more critical need to provide for *each* of our customers. Over the years, we had learned how to customize but, for the most part, had not learned to *personalize*—to put a face on each faceless customer.

Even today, as often as we claim that we are customer-oriented or customer-driven, all but the very best remain *customers*-driven. Indeed, flying in the face of good sense and past experience, most businesses are *still* trying to convince their customers that the one size—or, for that matter, the 20 sizes—they offer is the size the customer wants rather than tailoring the relationship specifically for each customer. Most companies, in other words, are *talking* One Size Fits One at the same time that they are trying just about everything to keep customers coming back—*except* to give them what they want and to treat them the way they want to be treated.

In a world where customers have the power, as long as we continue to convey that we are more interested in getting than giving and that we are more concerned with creating an impression with customers than with caring about them, we will never win the loyalty we seek. And the only way we will win that loyalty is by demonstrating our caring in an irrefutable manner—by serving every customer in a way that lets the customer know that he or she is understood, by giving each of them the special treatment that creates an emotional as well as a rational bond. Only then will the relationship have the sufficient depth and meaning to endure over time.

Simply put, most of us have continually underestimated how much we'll have to change—and keep changing—to build loyal relationships with our customers. We wanted to believe that a satisfied customer would be a loyal customer, and we were disappointed. Often, we spent more time thinking of what's in it for us than we did thinking about radically improving the value that we deliver to customers. Many of us held onto the past for far too long. We made incremental changes and "hoped" that these changes would lead to dramatic differences in the nature of our customer relationships. At times, we assumed that recent service improvements would overcome the inherent distrust that had evolved in most commercial relationships.

In short, we didn't comprehend the magnitude of the challenge. And, meanwhile, the best of businesses among us have begun to discover what—as customers ourselves—we knew all along: that we are not easily impressed. and that we certainly know the difference when somebody genuinely cares about us and when somebody is trying to manipulate us into thinking they do.

Today, a growing number of end users are *already* making the choice to give their business only to those companies that will listen to them, will build a strong relationship, and will provide them with the special value they are seeking in exchange for their dollars. As a result, we have little time to waste. Now's the time for us to make a choice as well: not about how we can shoehorn customers into buying what we have to sell but, as One Size Fits One becomes an everyday reality, about the kind of company *ours* will be.

Will we continue to look for more customers who fit *our* idea of what they need and for more sophisticated ways to manipulate them into being loyal? Or will we move toward an altogether new relationship culture—one governed by a value equation that balances the rational *and* the emotional, that is driven by caring, and is focused on creating customer benefit?

In most cases, making the choice for this new relationship culture will not be for the faint of heart. If we think that the choices that many of us have already made are significant—hiring customer service representatives, creating help desks, liberalizing return policies, performing costly surveys and follow-ups—they pale in comparison to the choices that must still be made. There is, in fact, a good chance that most companies have chosen not even to *see* the real choices because they are so uncomfortable, so revolutionary, because they fly directly in the face of "business as usual."

How so? Involved in the choice of entering into the One Size Fits One world will be the creation of an environment where employee bribes for compliance are replaced by employee commitment; where competition exists in the marketplace, not in the hallways; where continuous improvement exists side by side with significant innovation; where processes are improved instead of task-managed; where change is embraced rather than undermined; where information is gathered by all and treated on a right-to-know basis rather than as a way of hoarding power; and where customers and employees work together to establish mutually created value.

In short, choosing to care about each individual customer will require that we reshape our thinking about how and why we are in business and how we provide the products and services we do—whether our business is large or small. This choice may well also require a major reordering of our priorities: Will we, for example, persist in the unassailable primacy of shareholder return over true customer satisfaction? Most significantly, the choice to enter the One Size Fits One world will demand a huge leap of faith from the belief that customers must be manipulated to build loyalty to the belief that *caring* must come first and that loyalty and profit will follow.

In making this choice, we offer one overriding caution: To be tentative about your decision in today's warp-speed marketplace is tantamount to not making the decision at all. For there is little doubt that others *will* choose in favor of revolution, not evolution, and the rewards will go to those who build relationships with the most profitable customers first.

Fortunately, beyond the significant challenges and tough going we confront in making this choice, and the changes that inevitably will accompany it, lie enormous hope and opportunity. We, as customers, will finally be able to get more of what we want for our dollars. And companies that can first mobilize, and then capitalize on, the enormous pent-up demand for the personalized service and caring that most customers have for decades been yearning for will prosper.

By far the greatest hope and promise of the One Size Fits One world, however, lie ahead for all of us as employees—whether we work on the front line or in the executive suite. Tools such as traditional performance appraisals that force us to compete with our peers and restrictive, mind-numbing job descriptions that thwart our ability to care and to do the best job we can for customers will disappear. They'll be replaced by more enlightened tools and systems that encourage our intelligence, give credence to our creativity, and enable us to build the relationships with our customers that our companies have to have. When this happens—*and it will*—the essential nature of the workplace will change.

Where today, for many of us, the most hopeful day on the job is our first—before so-called reality sets in—in the new relationship-based world, every day will hold that hope. For the only way individuals can provide a One Size Fits One experience for their customers is to work in an intrinsically rewarding and satisfying environment that gives them the freedom to exercise the most admirable and noble elements of the human character.

Sure, it's easy to be cynical and say that hope and work, caring and profit, or jobs and pleasure don't mix. But, for those companies that make the choice to embrace the One Size Fits One model, these concepts will meld in an exciting, new, relationship-based culture that will be rewarding and fulfilling for us all.

PART TWO

Building Relationships One Customer at a Time

There is only one boss: the customer. And he can fire everybody in the company from the chairman on down, simply by spending his money somewhere else.

—Sam Walton, founder, Wal-Mart

THE BUCK STARTS HERE...

Since the customer will dramatically influence the economic relationship in the future, Part II of this book has been written in the customer's voice rather than the company's or an outside source speaking for the company's interests. Not that the customer is always right or that customers always know what's best for them. As both customers and businesspeople, we're well aware that customers can be wrong and that, frequently, those they do business with even know more about what's good for them than they do.

Nevertheless, after years of having their voices muted, we believe that it's time to hear the customer out, loud and clear. And that, as we move into the One Size Fits One future, if we as businesses err, it should be on the side of listening too closely or too much rather than too little.

Specifically, we have chosen the voice of the end-user customer, the customer whose dollars drive the entire process. The buck starts here, in other words, and its pull is felt from internal department to internal department, strategic partner to strategic partner, and person to person, all the way up the distribution chain and down again.

No one, therefore, can afford *not* to understand who this end-user customer is and what his or her wants are. For it is only from this understanding that a successful collaboration among strategic partners to create value can emerge. And it is only through this understanding that a successful long-term relationship between the final service provider and the ultimate customer can develop.

That's why, when you read through this section, you'll see a preponderance of examples about the consumers of products and services rather than examples of how partners within the process might cooperate to make the delivery of value possible. We also recommend that you reconsider before you say, "That's not us—we're no 7-11; we build parts for mainframe computers," or, "We're doing actuarial work in a large insurance company," or "Our customers are large, sophisticated companies; we're different."

Ultimately, whatever we do will affect the value inherent in the product or service that the end-user customer buys, whether he or she is the CIO of a Fortune 500 company trying to integrate multimillion dollar systems or a concerned spouse taking out a disability insurance policy to provide for family members in case of an accident or illness. And all the value that is added along the distribution chain will come to naught if that end user is not the primary focus of the value creation process.

REMEMBER US?

Curious things are happening. More and more companies, it seems, are trying to turn us into loyal customers. We read about it in the papers and the newsmagazines—in *Time* and *Newsweek* as well as *Inc.* and *Fortune.* We hear about it on the radio and on TV. Companies in every industry are saying they want us around for the long term. They say their future success actually depends on keeping us in the fold. Executives proclaim that the only profitable customer is a loyal one and that the only way to grow a good business is by building good relationships.

Just about everywhere we look, in fact, there's evidence of our newly discovered importance to companies we've been doing business with for years. We go to our bank, our car wash, or the grocery store, and everyone is wearing "The Customer Is #1" buttons. There's a "Customer Credo" next to the register at our local minimart. Our insurance agent has a "Year of the Customer" banner in his office that's been there for at least three years. And all of us who work anywhere have T-shirts, sweatshirts, caps, umbrellas, pens, or beach towels with "customer-this" or "customer-that" printed on them.

And, if all this attention to customers weren't curious enough, we're actually beginning to believe that some of the companies we deal with are genuinely serious about keeping us as loyal, happy campers.

But do businesses really understand what it will take to win our loyalty? To come back to them again and again? To buy everything we buy from *them?* To share information about ourselves with them? To pay a premium for excellent service and then tell stories about it? To try their new products? To send our friends? In short, to *trust* them?

AVOIDING THE "FREE DESSERT" SYNDROME

The truth is, it depends. It's different for every customer. It varies according to our individual needs and expectations, *what* we like and what we don't, *who* we like and who we don't. In other words, there's no generalizing about what makes an individual or an individual business loyal.

Still, there *are* a few things we do know about what makes us loyal—if not from our business relationships, then from our personal ones. We *hate* being taken for granted. We don't want to be treated as if we have no other options because we do. We can take our business elsewhere or drop it altogether.

America's Pastime—Still Striking Out

While new baseball stadiums open to rave reviews in cities like Cleveland and Baltimore, and as yesterday's simple ballpark hot dog is replaced by a gourmet selection of sausages and more, Major League Baseball has yet to recover from the strike that cut short the 1994 season and resulted in the cancellation of the World Series for the first time since 1904. Chances for a full recovery, in fact, are slim...given MLB's proclivity for shooting itself in the foot.

The 234-day strike, which began August 12,1994, and continued until March 31, 1995, was the eighth work stoppage since 1972. While players and owners, consumed by their own interests, wrangled in court and at the negotiating table, fans fumed.

If fans are sheep—as MLB may have hoped they were—then they're still flocking...but not, with rare exceptions, to major league ballparks. During the strike, the Detroit Tigers sold less than half of the 52,000 tickets for Opening Day, normally a sellout event. Attendance at MLB games in 1995 was down by 20%, and league revenues were 22% below those of 1993. Television advertising revenues plunged as local sponsors stayed away. The Houston Astros managed to keep some of its major accounts, but only by offering a 50% discount.

According to a survey commissioned in February 1996 by the owners' association and conducted by Penn & Schoen, more than one-third of respondents said MLB was not committed to the fans.

Just as they do in the yearly All Star balloting, America's fans are voting. But this time with their feet, not their pencils. And while some enterprising club owners are enticing fans back with $1 tickets and free, in-your-stadium-seat cellular phone messaging and calling services, the successful debut in 1994 of the customer-focused Texas-Louisiana League and the arrival of the United Baseball League in 1996 have further diminished Major League Baseball's hold on the American imagination.

Could it be that if things don't change soon, MLB's once exclusive field of dreams may slowly turn into a nightmare?

Sources: Aaron Bernstein, "Baseball Owners: 1, Players 0," *Business Week,* April 17, 1995; Keith Hammonds, "Okay, Baseball, You've Got One Last Chance," *BusinessWeek,* April 1, 1996.

We like it when businesses listen to us. We want to be respected. We want to feel that we're important to the people we're dealing with and that they genuinely care about us. We want to feel special—which is to say that, in a competitive environment, the company that merely *satisfies* isn't doing the job; satisfied customers are not necessarily loyal customers.

The way we see it—and this goes right to the heart of the issue—it's unlikely in this day and age that *any* company, no matter how low its prices, will win our business over the long term with everyday, or ho-hum, one-size-fits-all products or services.

Quite the contrary. As the future unfolds, it's becoming increasingly clear that, as customers, before we give our individual loyalty to any company, we're going to demand what we've wanted for years but until recently had almost given up on: to be treated as individuals. That's right, as in One Size Fits One. Each customer. Every *one* of us. In a One Size Fits One relationship.

It's how we feel we're being treated. For example, suppose you call a hotel to book a nonsmoking room. When you show up, suitcase in hand, a front-desk clerk explains that he's sorry but there are no nonsmoking rooms available. Compare that with being told, at the time of the reservation, that no nonsmoking rooms are available but that the hotel will call you and let you know if one opens up unexpectedly. In the end, you might not get a nonsmoking room, but you'll at least feel that you've been dealt with in an honest fashion.

Look around at products and services that have passionate fans. To be sure, those products and services must be high quality. But are they really that much different from their competitors? Probably not. The difference lies in the relationship between the company and the customer. Harley-Davidson enthusiasts don't just love their bikes; they love being part of the Harley family, and they'll gladly spend their own money to travel to Harley gatherings to meet kindred spirits.

Similarly, Saturn owners drive from all over the United States to congregate at the plant in Spring Hill, Tennessee, to swap stories, tour the assembly line, and meet with Saturn's corporate staff. Sound crazy? It shouldn't.

When we hear about emotional attachments like these, we're tempted to wonder *why* any business, large or small, persists in treating the human dimension as an afterthought or, even worse, a bother. Ask yourself: When you've purchased a product that doesn't work, or doesn't work the way you thought it would, which is more aggravating: not having the product you really want in hand or feeling as though the business from which you bought it has taken your money and run away?

Power Shift

A gentleman we'll call J.D. went into a branch of a nationwide coffee company to purchase a cappuccino maker for a friend. After paying $600 for a machine, J.D. opened the box to find it dented and rusted.

Instead of taking the cappuccino maker back and replacing it, the management at the store questioned whether J.D. had caused the damage.

J.D. tried to take his complaint to the upper levels of the company management. He was met with condescension, lack of interest, and even stonewalling.

J.D. was determined not to let the incident go unnoticed. He used $10,000 of his own money and launched a national campaign (complete with a toll-free telephone number) against the coffee company. A full-page ad in a national newspaper recounted the whole story, blow by blow.

When we caught up with J.D., he had just filmed a segment for the television program *Hard Copy*, had been interviewed by the *Wall Street Journal* and the *New York Times,* had been featured in futurist Faith Popcorn's newest book, and had appeared on a national call-in radio show. Rather nonchalant about the intense media attention, he wanted to talk more about why he was so disgusted with the coffee company.

He told us that he would rather not have spent the money or the time, but the coffee company didn't seem to care about his complaint or the personal embarrassment the unsatisfactory product had caused him.

What happened? Does it matter? He bought a product and he wasn't satisfied with the product or the company's response to his complaint. He felt they didn't care. And so, the rest is history.

Is J.D. a one-in-a-million rebel with a cause, or does he represent the future—the growing number of activist customers ready to take action when the status quo is not to their liking? You make the call.

A second reason we're being so stingy with our loyalty is that an enormous number of choices are suddenly open to us. Everywhere we look, it seems, we're inundated with new choices, more choices, and better choices. Huge retailers like Wal-Mart, Price Costco, and Target are pleading their cases for more of our money. Not satisfied with all they've already been selling us, they also want to be our grocer, our pharmacist, our optometrist, our stationer, and our auto repair shop. Circuit City, where only yesterday we bought our tape recorders, electronic gear, TVs, and washing machines, has now gone into the used-car business big-time with CarMax.

Used Cars: Can We Be in the Driver's Seat?

Jokes about used-car lots and used-car salespeople are a staple of popular humor. You know—sleazy salespeople with Cheshire cat smiles trying to pass off junk, cars that fall apart the moment you start the engine, pricing that's both incoherent and insulting.

What bothers us is not used cars per se, but the way we have to buy them. We feel like we've been taken for a ride.

Enter CarMax and perhaps we view the future in the re-invention of used-car sales. The auto superstores owned by Circuit City sell used cars through no-haggle pricing, mammoth selection, and time-saving technology.

CarMax offers prospective buyers a browse through available cars from computer kiosks that print photos, prices, and specifications of the cars selected by the customer.

While the customer takes a test drive, the computer goes back to work, completing a credit check and paperwork for the auto loan. Since CarMax is connected to the state's Department of Motor Vehicle data bank, tags, title certificates, and license plates can be handled on the spot.

When the customer returns from the test drive, CarMax mechanics provide a written offer for the trade-in car, which is good whether the customer buys a CarMax car or not. The entire process takes less than an hour. If the customer buys the car, he or she leaves the lot with the car, the tags, the registration, and a 30-day guarantee with extended warranties available.

Source: Motley Fool's Today's Pitch: Motley Fool@aol.com and "The Retailing Revolution: How Superstores and New Technologies Are Changing The Way Cars Are Sold," *International Automobile Dealer,* Vol. 13, No. 3. May/June 1996.

Years ago, our choices were limited, and demand far outstripped supply. But the growth of the consumer-oriented economy has resulted in a deluge of goods and services. Looking for athletic shoes? Name your sport and then pick from at least 20 brands. Ice beer? In 1993, we had a single domestic brand to choose from; in 1994, there were 36. We open our Yellow Pages and find as many pizza parlors as lawyers. Crack open the business pages and we find as many mutual funds as individual stocks. Bottled water? Restaurants? Microbreweries?

The list goes on and on. We read in the paper that, in 1981, there were 2,689 new products introduced on grocery and drugstore shelves; by 1991, that number had skyrocketed to 16,143.

With so many companies out to win more of our dollars by offering us more choices, it's actually gotten to the point where it's hard to tell *what* business some of these companies are in. Is the Hard Rock Cafe in the food business, the entertainment business, the memorabilia business, the T-shirt business? When we visit the great Mall of America in Bloomington, Minnesota, are we going to a shopping center, an amusement park, or a destination resort? How about when we call Lands' End? Are we ordering from a clothier, a luggage boutique, or a dry goods emporium? Everyone, it seems, is trying to sell us everything under the sun. And we've got the choices today to prove it.

WE WANT IT OUR WAY

A final reason for our reluctance to give our loyalty is that, as a group, consumers face a historic moment. Until World War II, consumers accepted whatever industry produced. Supply was not linked to demand, and companies clearly had the upper hand in dictating consumer choices.

After the war, pent-up consumer demand led to massive production, and the 1950s and 1960s were times of almost insatiable customer desire. We wanted—and took—all the cars that Detroit could make. We wanted refrigerators, television sets, homes, trailers, everything.

But a power shift was in the making. The nascent consumer movement of the 1960s showed people that they do have a voice in the marketplace. Overcapacity resulted in swollen inventories, adding to customers' power. The growth in competition from other nations put pressure on domestic industries to improve their products and services. Suddenly, we could reach out and choose from all over the world. And inflation and stagnant incomes made people even choosier and more determined then ever to get their money's worth.

We've also got more information than ever before. Thanks to the radio, TV, print media, and the Internet, we can compare and contrast goods and services as never before. We're also growing bolder about voicing our discontents, and we're finding that companies *do* listen. We've never been sheep; it's just that we've been treated as though we were.

Stamping Our Virtual Feet

Cyberspace is uniquely suitable as a forum for consumer discontent. With the click of a button, a person can broadcast a message to thousands, even millions. Companies are beginning to recognize that bad publicity on the Web isn't a force to be ignored:

- Two British scientists seeking to force a fast-food company into changing its nutritional and environmental policies launched BurgerAttack, a World Wide Web page devoted to publicizing their cause. After only 16 days on the information superhighway, the page was visited by 135,000 people throughout the world.
- The Angry Organization (http://www.angry.org), billed as the most feared home page on the Net, asks people to send in their complaints about organizations, products, company policies, and the like. With over a half million visitors in six months, it is growing in popularity.
- Motley Fool on America Online (Motley Fool@aol.com) is a personal finance and investors' chat forum. At least three times in 1996, the chatter among share-

holders, employees of the companies mentioned, and investors managed to send the stock of targeted companies crashing —in a matter of minutes. In one particular case, a visitor to the site with the screen name Cheeseboyl (in real life, a policeman from Woodbridge, New Jersey) criticized a high-technology stock as overvalued and encouraged fellow investors to sell. After weeks of reading Cheeseboyl's tirades, the chairman of the company added his own posting, extending an invitation to Cheeseboyl to come to the corporate headquarters to talk. When queried, the chairman said that it had become clear to him that the situation was not going to go away or be silenced, so he decided to take a proactive approach.

- When corporate raider Paul Kazarian lost the battle to buy Borden, Inc., he launched the world's first cyberspace corporate takeover site (http://www.japonica.com). From this Web page, Kazarian gathers intelligence, conducts research, and provides sophisticated analysis for investors. It's also a base from which he amasses the e-mail addresses of directors and shareholders of companies that catch his eye. Kazarian offers a service that enables site visitors to send public or private messages to institutional investors, the press, and corporate directors around the world.

Thanks to the mass media and tools like the Internet, we've also learned that a growing number of "vigilante customers" in our midst are taking on the corporate powers that be—and winning—just as futurist Faith Popcorn predicted we would, and that all around the country these vigilante brethren of ours are alive and well and actively recruiting. Thirty years ago, Ralph Nader and his band of young, idealistic followers, "Nader's Raiders," were just about the only consumer advocates in sight. Actress Betty

Furness went from pitching Westinghouse refrigerators to commissioner of New York City's Department of Consumer Affairs. Today's activist customer is as likely to be the formerly mild-mannered, suburban businessman, our neighbor, or even one of us.

The question we more moderate customers have is: Short of setting our car on fire in front of our local dealership to protest shoddy service or taking our mailman hostage because we're frustrated with the U.S. Postal Service, what can we do to get more and better results from those with whom we do business? Simple. Many of us are doing it already. We're hitting companies where we know it hurts them the most—in the pocketbook. And we're doing it by *voting,* first with our wallets and then with our feet.

We've forced a major soft drink company to reinstate its old formula. A national consumer goods retailer, dissatisfied with one of its key supplier's billing and invoicing procedures, threatened to pull all its business if the supplier didn't change its ways. Consumer pressure led the U.S. Food and Drug Administration to revamp nutritional labeling. And widespread disgust with sloppy environmental practices after the Exxon Valdez spill was a factor in the decision by major oil companies to use only double-hulled tankers.

MEET THE NEW CONSUMER

As a result of companies' scrambling for our loyalty, coupled with our long-running frustration with their ways of doing business and our ability and willingness to take our business elsewhere, we're gradually getting a greater and greater voice in how companies treat us. It's only a start, but companies are starting to listen to our concerns about what we get from them and how we get it.

Instead of ignoring us as they used to, or reciting their traditional one-size-fits-all policies chapter and verse, companies are *actually starting to customize their rules and to alter their ways of doing business to win us over and keep us coming back.*

The power to call the shots in the marketplace is beginning slowly but inexorably to shift from the organizations and individuals we do business with to us. And it doesn't seem to matter if we're buying a washing machine for our home, a pizza for our kids, or a paper shredder for the office. Nowadays, we're expecting service and services that didn't exist even a year ago and asking for things we've never dared ask for before. And more often then not, we're getting them—or soon will be.

The Nordstromizing of America

Have you ever seen a car with a license plate holder that says, "I'd rather be shopping at Nordstrom"? Think about it: *People actually pay for the privilege of carrying advertising.* What kind of a store inspires that kind of passionate loyalty?

Nordstrom carries many of the same brands that other national retailers do. Its stores are laid out in much the same way as other large department stores. Many malls and shopping centers feature a Nordstrom store and a competitors. What, then, makes Nordstrom so different?

The key is Nordstrom's service. Ask any fan and she or he will tell you about a favorite salesperson—one who calls up to let you know that something in your size has just come in, or takes the time to assemble whole wardrobes.

Any department store can sell clothes. But, at Nordstrom, the actual selling seems almost incidental. Its real business is *forming intimate relationships with customers.* A Nordstrom salesperson doesn't simply ring up a sale; he or she is a *relationship manager,* taking the time to know and understand each and every customer.

Nordstrom's level of customer commitment is quickly becoming a benchmark. No wonder leaders continually tell us that they want to be the "Nordstrom" of their respective industries. A large retailer proclaims that its goal is to become the "Nordstrom" of self-serve stores. The facilities division of a large defense contractor desires to be the "Nordstrom" of defense contractors. Our favorite was the quote in the *Los Angeles Times* in which a member of the City Council said that the Los Angeles government would become "the Nordstrom of city government."

There are examples in English of a proper noun becoming an adjective to describe a particular type of service. For example, the term FedEx doesn't necessarily refer to just the Memphis-based company—it's shorthand for generic overnight delivery. People say they're going to Xerox a document when they actually mean photocopy. Today, there is Nordstrom service—no other description is necessary.

Five years back, for example, a car owner facing a major repair was simply out of luck if the warranty period had expired. The owner wouldn't even *think* of calling the manufacturer. Today, chances are that a complaint lodged with the dealer who sold the car would result in a free repair with a minimum of hassle—especially if the car owner is a good customer who's been bringing the car in for service since it was new. And, five years ago, a company unhappy with its $30,000 photocopier was stuck with it. Today, if that same company were to buy a machine from Xerox, the faulty unit would be whisked away and the money refunded. The only question Xerox might ask is, "What can we do to make it right?

Of course, knowing what it takes to impress a customer was hardly pioneered by Xerox. There have always been a variety of businesses that have been ready, willing, and able to earn customers' loyalty.

But most of these businesses have been small and local: our dry cleaner, for example, who'll do whatever it takes to make sure a suit is ready in time for a business trip—including hand-delivering it after business hours to our door. The restaurant down the street whose host not only remembers our name but always manages to find us a table, even on the busiest evenings. The corner stationery store that understands if we forget our wallet and trusts us to pay tomorrow.

But, recently, a number of larger companies—such as Xerox, Southwest Airlines, USAA Insurance Company, Ritz-Carlton Hotels, American Express, MBNA Bank, and many others—are starting to turn things around by taking the first steps necessary to win our long-term commitment. Despite their size, these companies are finding ways to "act small" and give us as much special attention as... well, our local dry cleaners.

The Dog Ate My Microsoft Word. . .

Writing a book with the latest technology is a vast improvement over banging away on the keys of an old Underwood typewriter, unless you experience a hard-drive crash on your computer. That's precisely what happened to us. After nightmarish weeks of trying to recover data and disks and rebuild bytes, it was time to reload all the versions of the software programs. Lo and behold, it was then we discovered that the least organized of us, while cleaning out his garage, had thrown away the Microsoft Word 5.0 program and the newly purchased Microsoft Office upgrade. Hundreds of dollars of software gone and a manuscript deadline prompted a desperate call to Microsoft.

We began by telling the person who answered, "This has to be true. Who could make this stuff up? We know it sounds like the old 'the dog ate my homework,' but it wasn't the dog, it was really my boss! We have the receipts for the software; we can fax them to you. We can even send you the documentation. Is there anything you can do?"

The Microsoft customer service representative listened to us without interrupting, interrogating, or even requesting that we send our firstborn for an even trade. Then she did the most amazing thing—she offered to send us a replacement free of charge!

Needless to say, there is one among us who will remain a Microsoft Word user forever. Microsoft got itself an apostle, who retells the story a dozen times to anyone who will listen—a walking, talking advertisement for the company, its service, and its caring.

And, in the end, we never took delivery of the free software. The least organized of us found the copies of the software programs—the day after our call to the software giant—in his brother-in-law's garage!

A MOVEMENT TAKING HOLD

The fact is that a growing number of businesses, large and small, are beginning to catch on to what we're looking for as customers. Service is definitely on the upswing. Products are far more reliable. Guarantees are longer. Warranties are more comprehensive. Companies in most every industry seem to be making at least a halfhearted, if not a genuine effort, at providing us with better value and a more satisfying buying experience.

For example, it used to be that the standard 10 A.M. to 3 P.M. banking hours were, putting it mildly, very inconvenient. Today, with telephone and on-line banking and easy-to-use ATMs, we can get cash, speak to a banker, check our balances, and perform many other complex transactions at any hour of the day or night.

If You Can Dream It, They'll Deliver It

It used to be that only dry cleaners, pizza joints, grocery stores, and florists delivered. Today, almost every type of product is being offered for delivery right to our door, 24 hours a day, 7 days a week.

A video rental store will deliver the latest movie to our home and pick it up the next day. A 24-hour Walgreen's drugstore will deliver a box of diapers and a pacifier in the wee hours of the morning. You can have fresh organic fruits and vegetables picked in the morning and delivered to your door that afternoon, along with a supply of assorted fruit juices and milk from Full Belly Farms.

A service called Pet Kabs can have your guinea pig delivered to the veterinarian in the morning and returned to you that evening and they even customize service depending on the animal and its preferences. (One client owns a Great Dane that won't ride in vans and is terrified of garbage trucks. Pet Kabs rents a limousine and picks up the dog at 5 A.M., before garbage trucks hit the road.)

You can even have your children delivered to their schools, day care centers, or the local Little League practice by Kids Kabs. Just think what tomorrow will bring!

The slogan of Federal Express is: "The most important package we ship is *yours.*" It's possible to track the progress of your shipment by telephone, or through a PC with free software provided by FedEx, as is the case with many of the overnight delivery companies.

We call Lands' End to send a gift to a friend, and the service representative taking the order reminds us that we sent the same gift to that person two years before.

We pick up the phone in Duluth to order a pizza from a Pizza Hut, and the person on the other end of the line asks if we'd like the same ham-pineapple-sausage-mushroom-hold-the-jalapeños combo we'd ordered a week before in San Diego.

About Last Night...

The hypoallergenic pillows we requested during our last stay are on the bed, all fluffed up—and we forgot to ask this time. There are numerous extra towels (and we remember we had called room service for extras during our last visit). The cookies on the tray are all chocolate chip, our favorite kind—and the oatmeal ones we received last time but didn't eat are mysteriously missing. When we checked in, the concierge asked us if we wanted tickets to the symphony as we had requested last time. We begin to realize that the Ritz-Carlton has taken *every bit* of information it learned about us from our last visit and indexed it in a database. Before our arrival, the hotel staff, from room service staff to the chambermaid, customized our room with the extra touches they knew we would want or need. They seem to know us as individuals and they seem to care genuinely whether our stay is enjoyable.

If we're in the market for a pager, Motorola gives us more options to choose from than we had when we bought our last car—and, unlike our auto purchase, there isn't a complete "performance package" we have to order to get the single option or options we want.

We walk into our local clothing store, and our salesman has a special chapter with our name on it in his "customer book." In the old days, we'd have been pleased if he recognized us when we walked through the door. Today—for better or worse—he's able to recount the size, color, style, and price of everything we've ever bought from him or anyone else in the store.

And it doesn't stop there. Today, Marriott Hotels offer the 10-second check-in—we give our name to the bellman at the door and he hands us our key. Peapod Interactive Services lets us order groceries and arrange for delivery, all on our computer. With Waiters-on-Wheels, we can get the specialty of the house from some of the best restaurants in town—along with virtually everything else on the menu—served to us at our house. We send for a Country Store catalog and we can order custom-sized quilts and bedding. The custom-made bikini and custom-made Levis are already here, and custom-made Nike running shoes are on the way.

Meanwhile, other companies are doing their best to win our loyalty with coupons, discounts, free nights, and other assorted add-ons and goodies. Still others are inviting us to join their frequent-flier, frequent-guest, frequent-shopper, frequent-skier, and frequent-just-about-everything-else clubs. Even those notoriously poor service companies we used to kid about—the ones we thought would never get it—are showing signs of life.

Ox •y •mo •ron (ak' si mor' an): a figure of speech in which contradictory terms are combined as in: Law Firm Guarantees Service or Your Money Back.

The Chicago law firm of Coffield Ungaretti & Harris offers clients a written guarantee that reads, "We cannot guarantee outcomes; we do guarantee your satisfaction with our service or your money back." That service consists of "cost-effective" and timely service in which the client will be consistently kept informed of progress.

Although the idea of a service guarantee is not new, its application to a law firm may be a first. The guarantee was developed as the next logical step in client service, according to partner Richard Ungaretti. He notes, "It shows we're committed to clients, and it makes clear to everyone in the firm that we are in the service business."

American Bar Association Journal, Reske, Henry, Vol. 81, 8-1-95, p. 18.

The fact is, with few exceptions there's hardly a company out there that hasn't shown *some* improvement in dealing with its customers over the last few years. Even the most stubborn, wrongheaded companies have inched in our direction. Either that, or they've gone out of business.

Unfortunately, most companies still don't give their frontline employees the power to do what's needed to form One Size Fits One relationships. There's a particular frustration we feel when the bonds we form with the people of a particular local store or branch office are wrecked from above by rigid corporate policies. We know and like our local bank managers, opticians, pharmacists, and insurance agents. But their ability to serve us, to form One Size Fits One business relationships, is more often than not thwarted by one-size-fits-all rules and regulations handed down from headquarters. They know it and we know it.

Maybe Some Relationships Are Forever

Large manufacturers are finding new and creative ways to nurture relationships with customers in the hope of broadening that relationship over time (increasing their share of wallet) and keeping that customer in the fold "for life." By adding a host of new services to the products they sell, these companies are putting themselves on the customer team. In addition to after sales service, maintenance, and technical training, many companies have added services that range from strategic planning, consultation and quality improvement training to leadership education—and it's paying dividends.

By some estimates, General Electric garners nearly 60% of its profits from services (up from 16.4% in 1980) and margins typically run 50% higher on services than on product sales. At Otis Elevator, two thirds of its revenues come from service and maintenance. Companies like Baldrige Award winner Xerox and Deming Award winner Florida Power and Light have long sold education based on their improvement methodologies.

What we are witnessing today is only the beginning. As we learn to manage processes across organizational boundaries, the lines between products and services will blur even further. Retailers will be the front end of the manufacturing process and manufacturers will help to manage the sales and service processes as we learn to manage the value creation process more effectively.

The truth is, the local level is a good place for businesses to start because it's the front-line employees who get to know customers, hear their complaints and their compliments, and see the actual effects of corporate policies. So the business may have uniform policies coast to coast, throughout hundreds or even thousands of branches, but does it have loyal customers?

Service Worth The Paper Its Printed On

Last year we spent several hours talking with small business owners about how the best manufacturers were quickly becoming service companies. After the discussion, we were met by at least 20 people representing different businesses who wanted to talk about a supplier. What was surprising, given the diversity of the group was that it was the same supplier, in each case, Appleton Paper.

Each of the people described how the people of Appleton Paper helped them better understand their customers' needs. Appleton Paper even assisted them in devising new ways of building better relationships with their customers. They described the service training and the continuous improvement education that Appleton provided their staffs. Each person had a different story to tell, but the bottom line was always the same—they appreciated the fact that Appleton was dedicated to making them successful.

Partner with suppliers—the words are common these days. What we saw in the eyes of these customers that afternoon was anything but common. To them partnership and trust were spelled A-P-P-L-E-T-O-N. Funny thing—we never talked about the paper!

WHAT'S THE NEXT STEP?

So, if this is the case—with the U.S. Postal Service finally accepting our Visa and MasterCards—why is it that most of the companies we deal with still seem to underestimate what it takes to win our loyalty? This is despite the fact that, from their perspective, they *seem* to be providing us with the One Size Fits One products and services we've been asking for.

The question is, *are they?* Are they really? Or are most businesses today hooked on thinking of One Size Fits One primarily in terms of using "gee-whiz" technology to customize a life insurance policy or monogram a sweater? In other words, *instead* of building the complicated relationships that lie at the heart of winning our loyalty, are they opting for the quick fix?

Do these businesses truly believe that the *depth* of the relationships they form with us, and not merely the *number* of them, will be the key to their future success or failure? Are these businesses even aware of how extensively they will have to change to make these relationships work? Probably not. Not from what we've seen. Not really.

For example, do they know that a loyal business relationship with us or our company is much like a relationship with any other human being—and just as complex? That it's more than valet parking, remembering our name, or flashing us a big smile every time we look their way? Just imagine if that were our modus operandi in the relationships we had with our friends—parking their cars, calling them by name, and giving them our pearly whites—that and "business as usual." How long would it be before they were out the door? We expect genuine commitment, sharing of information, emotional bonds, and the like from our relationships with family and friends. Why should we settle for less than that in our business relationships?

More to the point, do they understand that the loyal relationships they are seeking are *fundamentally different* from the relationships they've traditionally had with us? And, that to win our loyalty, it's going to take a whole lot more than they've already given?

The fact is, with the power in business relationships flowing in our direction, our loyalty can no longer be bought for a lower price, a dollars-off coupon, club "membership," or free miles—although we'll happily pay less, cash the coupon, avail ourselves of the advantages of the club, and convert those miles into a trip.

Rather, what we're looking for today in a business relationship is far more difficult to manufacture or come up with than customized products, services, and marketing ploys. What we're looking for in our business relationships is *companies and people who genuinely care as much about us as individuals as they do about their bottom line and whose actions bear this out.* In other words, it's the One Size Fits One relationship we're looking for, first and foremost, and not the One Size Fits One goodies that go along with it...at least not *just* the goodies. And until everyone—from the receptionist to the CEO—in the companies we do business with realizes this and acts on it, our loyalty will be hard to come by.

The good news for those companies seeking our loyalty is that we're genuinely on the lookout for companies, stores, business partners, and individual service providers to whom we can be loyal. Really. We like the familiarity, the good feeling, the camaraderie that comes from going back again and again to the same mechanic, the same salesperson, the same barber, or the same business vendor. Besides, it's convenient—not having to shop around always saves us time and hassle and almost always saves us money.

The bad news is that we're wary—both as individual customers and as business customers. Our response to many a company's sudden and intense interest in us is akin to a fear voiced by comedian Woody Allen: that the petroleum companies discover that human beings contain oil. As a consequence, before we commit to anyone, we want to be certain that his or her interest in an individual relationship with us is genuine. We want to know for a fact that a business really cares about us and is not simply trying to fool us into thinking it does. We want to know that we're truly number 1 in its eyes, not a distant number 2 to its profits and processes.

Adding to the bad news and standing squarely in the way of our giving loyalty is the past history we have with most companies—the frustrations we've felt, and frequently continue to feel, about the way they do business. Though a few companies genuinely seem to be "getting it" when it comes to creating customer-focused environments that encourage the types of relationships that will make us loyal, far too many merely claim to be getting it. They continue to make us go through all sorts of contortions to get what we want—even if it's just the most basic satisfaction.

It's as if these companies are in love with the *idea* of having loyal customers; they just haven't the faintest idea how to actually get them. Either that or all their talk about loyalty is just lip service; the words—and the pasted-on smiles—are there, but the spirit is missing. It's actually gotten to the point where we don't know what's more frustrating: those companies that just plain don't care about us and make no secret of it or those that are faking it by pretending to care but are going about their business as usual.

The truth is that, if these companies actually thought about the process, they'd probably find that they don't have to engage in massive reworking of their products and services. It's not necessary to burn down the store; what is necessary is a new approach, not new merchandise. In fact, if you're frantically trying to find out what's wrong with the goods? Then you haven't gotten the essence of the message. Customers probably aren't as mad at the merchandise as they are at *you*.

THINGS *ARE* LOOKING UP

There is definitely reason for hope. The marketplace is changing and every indication is that we, as customers, will be the major beneficiaries of those changes. In some cases, we already are. Many of us have recently been astounded and pleased by the service and selection at a newly opened superstore, bought a custom-made pair of Levis for only a few dollars more than an off-the-rack pair, and made a killing in mustard at our warehouse club. It's not uncommon to hear a friend say he was treated like a king when he took his car to the dealership for an oil change. A neighbor might express her surprise and delight at how quickly a long-distance provider adjusted an error in her bill. As the battle for our loyalty shifts into higher gear, things truly seem to be going our way.

How, then, will having it our way affect those with whom we do business? How much will they have to change to form the relationship with us that they say they're looking for? What precisely will they have to do? And what does *our way* even mean? As we said earlier, *our way* means something different for every customer.

Nevertheless, there *are* common elements in what individuals look for in a loyal relationship: being listened to, being respected, being cared about, feeling special. Translating these constants into a business context, we've come up with 10 things we'll be on the lookout for when we make our decision about who we'll be loyal to in tomorrow's world. We may not always have a clear view of it today, but it's a world where loyalty and commitment will be won one customer at a time and where one size no longer fits all or even a few but where One Size Fits One... and one alone.

Customer Rules for a One Size Fits One World

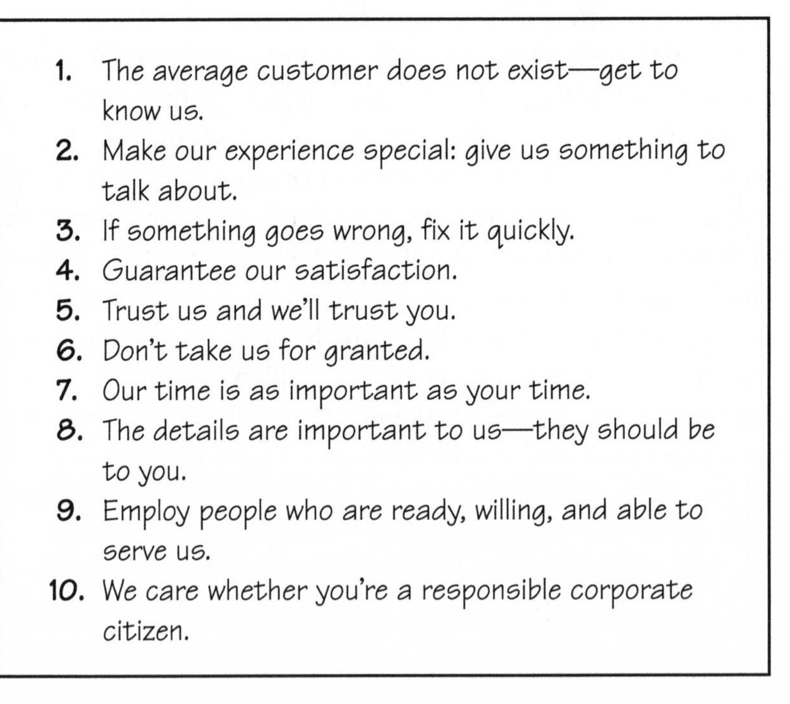

1. The average customer does not exist—get to know us.
2. Make our experience special: give us something to talk about.
3. If something goes wrong, fix it quickly.
4. Guarantee our satisfaction.
5. Trust us and we'll trust you.
6. Don't take us for granted.
7. Our time is as important as your time.
8. The details are important to us—they should be to you.
9. Employ people who are ready, willing, and able to serve us.
10. We care whether you're a responsible corporate citizen.

1.
The Average Customer Does Not Exist—Get to Know Us

...the marketer with the greatest scope of information about that particular customer...with the most extensive and intimate relationship...will be the more efficient competitor.

Don Peppers, Martha Rogers
The One To One Future

Let's get right to the point. There is no such thing as an average customer. We bristle at the thought of being considered average—in *any* of our relationships. And, in the years to come, with more and better businesses aggressively courting us for our loyalty, we'll be even less inclined to stay with a company that treats us pretty much the way it does every other customer. Averages factor out individual differences, and a company's ability to cater to those differences among customers will largely determine whether we'll stay or move on.

The fact is, thinking of us as your nameless, faceless, average customers is the antithesis of the relationship we're looking for. In the bad old days, our rough edges were worn smooth, our essences reduced to demographics. But, now, the information-gathering, management, and delivery tools necessary to know us as individuals are available at a cost that even the smallest business can afford. All that's missing, as far as we can tell, is the motivation and the inclination to use these tools. Of course, you want the data that will help you gain better access to our wallets, but we also want you to use the data constructively, intelligently, and intuitively to build a personal relationship with us from which we can both benefit.

So, as you're gearing up to win our long-term commitment please keep in mind that the average customer simply *does not exist.* We don't want to be your target audience unless we can be an audience of one. We don't want to be your niche market unless we're a market of one. And we never, ever want to be a demographic unless we can have it all to ourselves.

ANTI-ANONYMITY MEASURES

Knowing what we've told you about how we feel, you may want to consider the following six anti-anonymity measures:

1. Get to know us—each and every one of us.
2. All customers are not alike—recognize our differences.
3. Listen to us, but don't believe all you hear—the customer is not always right.
4. If you want information, ask us intelligent questions at convenient times—never ask us the same question twice.
5. Listen to the people who listen to us.
6. If you're going to walk in our shoes, then, walk in our shoes.

1. GET TO KNOW US—EACH AND EVERY ONE OF US

We don't doubt for a moment that a large number of businesses today are under the impression that they've made the choice to get to know us. The way we can tell is by the number of surveys and response cards we've been given and the quantity of information about ourselves, our families, our histories, our activities, and our incomes that we've conveyed by dutifully filling them out. We've been surveyed by the airlines, our rental car agency, a professional basketball team, our local newspaper, the hotel where we spent a night, the dealership where we had our car tuned, and the catalog company that sold us a sleeping bag. And that was just in the last month.

The problem, as we see it is that these surveys frequently miss the point—and, by extension, so do the businesses that pay to have them conducted. Just as Freud asked, "What do women want?" these ubiquitous questionnaires seek a one-size-fits-all answer to what we believe should be a One Size Fits One question. It's not what customers want that businesses today should be concerned with; it's what each customer wants. And, unless the results of these surveys are considered one customer at a time, rather than being toted up and averaged, they won't make a particularly effective tool to build better individual relationships with us.

We're sure that, in more than a few cases, the spirit behind these surveys is commendable and that the information they glean about us will be put to some future use for our collective benefit. For example, we've recently heard about a survey conducted by United Parcel Service that focused on the relative importance of on-time delivery to its customers. Before the survey, UPS assumed that, from a customer's perspective, timely delivery was the most critical factor in its operation.

The research revealed, however, that even more important than getting their packages quickly, many customers want information. They want UPS employees to have more time to spend to answer a wide variety of customer inquiries, including information about shipping rates and regulations.

As UPS changes its system to respond to these results, the survey may ultimately prove to have been a valuable tool for improving the value delivered to UPS customers. But, while this response might make more of us happier shippers, freeing up UPS clerks so that they can spend more time answering questions will not, in and of itself, guarantee the loyalty of a particular customer. It's how the clerks use that time *specific to customer needs* that will ultimately make the difference.

And then there are companies—again under the impression that they're getting to know us as individuals—that are more than missing the point with their information-gathering methods; they're actually frustrating us.

Why is it, for example, that we have to give the salesclerks at certain retail stores our name, address, and telephone number every time we buy anything at their store, whether it's a $2,000 computer or a 25-cent resistor, a $100 blouse or a $5 pair of nylons? Yet, when we return, the salespeople don't seem to have the faintest idea who we are. Nor do they even seem to care when we tell them that we've already given them that information and don't want to repeat it for the umpteenth time.

Most importantly, we still haven't the faintest idea why they want that information, what they do with it, and when, if ever, we'll benefit from having given it to them.

But I Didn't Call...

"I am loyal to USAA Insurance Company and have been for years. When I purchased my latest house, I was distressed when I was told that USAA would not insure the house because of its old wiring. I'm not sure how I convinced it to underwrite this policy, but I can assure you that groveling was involved. During the conversation, I told the company that I was going to undertake an extensive renovation within a couple of years.

"Fast forward two years and I'm up to my...ears in construction when the phone rings. It's USAA. Don't ask me how, but it remembered the conversation about the renovation. The person on the phone wanted to know if I had done it, because my insurance had not changed since I purchased the house. She could hear the construction noises in the background and we both laughed. She reminded me that it was no laughing matter—I was seriously underinsured—but told me not to worry; she would take care of me.

"And she did—even to the point of telling me that USAA's expert had valued our addition at an amount she thought might not be enough. She said that USAA would take my word for what insurance I needed.

"I was shocked. I hadn't asked. I had been negligent. I had dropped the ball. USAA doesn't have local agents. It had no way of knowing, except by listening and caring. And it did."

Don't get us wrong. It's not that we're reluctant to tell companies what they want to know. In most cases, we'd be happy to give you personal information—and far more than our names, addresses, and phone numbers—provided, of course, that it's evident to us that the information being used is for our *personal* benefit and isn't simply being thrown in your database hopper—or, worse yet, sold to someone else's.

To avoid having to ask each of us to provide our personal histories with each transaction, a number of companies have established "clubs" and issued cards to give their once faceless customers a distinct identity, albeit an electronic one. A clerk, cashier, or customer service representative can swipe the card (which has a magnetic strip or a UPC code) and immediately access pertinent information about us. And, unlike most survey results, this information is personal.

While this is a step in the right direction, the problem with membership in many of these so-called clubs, from our perspective, is that even with the information they have about us, many of the companies whose clubs we belong to don't do enough, if anything, with it.

Take any airline, for example. If we're enrolled in its frequent flier program, the airline knows where we fly, when we fly, and how often we fly. It knows how much we pay for tickets, how we pay for them, who we fly with, and where we like to sit. It even knows if we're vegetarians or on low-sodium diets.

But, even though it has all this information, and the additional data it's gathered from the surveys we've filled out, the airline has done little or nothing to make ticketing or check-in easier or to make flights more enjoyable.

Aside from being able to trade in our accumulated miles for an occasional free flight—replete with a bevy of restrictions—being a member of the club rarely makes us feel that we are special or that our relationship with the airline is important.

Jumping Ship

"Frustrated with my usual U.S. airline, with whom I've flown more than 3 million miles, I had my travel agent switch my reservation to British Airways. Within minutes of making the reservation, the travel agent got a call from BA asking: Who am I? How often do I fly? What would it take to win my business? What are my concerns?

"The very first BA employee I encountered at the airport greeted me by name and checked me in within seconds. Then, another BA employee asked if she could talk to me for a few moments so that the airline could serve me better in the future.

"After answering a few of my questions and telling me about BA's newest service innovations, she welcomed me to the family.

"This was more attention than I had received from my regular U.S. carrier in 10 years."

The key point to remember when gathering information about us is that when it comes to giving our loyalty—whether it's to an airline, a business partner, or a doctor—we're not looking for a data exchange but a *personal* exchange. We're looking, for example, for a store like K Barchetti Shops in the Pittsburgh area, where *every* salesperson in the store where you regularly shop knows who you are, what your tastes are, engages you in conversation, and genuinely seems to care about responding to your individual needs. It's as if everyone at K Barchetti Shops has taken on the collective responsibility to make your individual experience satisfying and pleasurable.

Not surprisingly, in K Barchetti and in stores and companies like it, our relationship is not so much with the entity but with the men and women who work there. They make the attempt to know us personally and deal with us as individuals. At Nordstrom stores, salespeople keep "personal books" on customers. That a business such as Nordstrom, hundreds of times the size of K Barchetti, does this, is a testament to the companywide attitude that places the customer first and to the famous retailer's understanding of the importance of using information to personalize service.

Maybe that's why we keep reading about Nordstrom. Its One Size Fits One treatment of its customers is a story that simply will not go away, particularly since there are new chapters written daily at Nordstrom stores around the country The same is true of the Ritz-Carlton, L.L. Bean, British Airways, MBNA Financial Services, and Bob Tosca Ford of Rhode Island. And are growing.

2. ALL CUSTOMERS ARE NOT ALIKE—RECOGNIZE OUR DIFFERENCES

While we essentially believe that how *much* we spend as a customer should not determine how *well* we're treated, we're not so naive that we don't understand that companies have to make money to stay in business and that some customers are more valuable to companies than others. We would, therefore, like to caution the businesses we patronize not to work overtime to create loyalty with *every* customer, particularly those that end up costing them more than the revenue or the referrals they'll ever generate. More to the point: We don't expect you to delight us if we have nothing or precious little to give you in return. Not all customers are alike. Not every customer is a potentially good customer. You don't have to knock yourself out for every customer.

Avoiding the BIG Mistake

Remember the movie *Pretty Woman?* Julia Roberts, dressed in her work clothes, goes into a Rodeo Drive boutique and asks for some sales help. The snooty manager looks her up and down and says that they have nothing she'd like and that she'd best leave.

The next day, Roberts, decked out in a smart new outfit and carrying enough clothing boxes to open her own boutique, stops in again and reminds the helmet-haired manager of her earlier rudeness. Roberts, who's carrying multimillionaire Richard Gere's credit card, cocks her head and says, "Remember me? BIG mistake!"

Mary Roe (her real name), a VP and branch manager of AMSouth Bank's busiest Florida branch, learned a similar lesson.

Two years ago, a plainly dressed older man walked into Roe's branch on a typically busy day. When he sat down in Mary's office, he began to talk about the recent performance of the Orlando Magic and the weather. Being polite, Roe didn't hurry him out, although she didn't expect him to become a customer—at least not a profitable one.

Pushing his chair back, the man declared that he liked the bank and wanted to open up an account. After filling out the paperwork, the man opened his checkbook and deposited $2.1 million. Today he's got over $7 million in Roe's bank.

Julia Robert's flashy clothes and the man's plain dress may have sent off similar signals—probably not worth the time. But Roe, unlike the boutique manager, profited by taking that extra moment to get to know her customer.

If all we ever buy from your convenience store is an occasional late-night quart of milk, we don't expect you to know our name or even whether we generally buy skim or the 2% variety; basic courtesy and the right change will be fine. And, if all we buy at your electronics store is a two-pack of AAA batteries, we don't want you to waste your time taking our name and phone number, even if it means we won't get your catalog. Trust us, we don't want it. And, if all we ask from your bank is a place to park an emergency $200, we don't expect the combination to the vault or even the preferred interest rate. In fact, the worst thing we could wish for you is a thousand more customers just like us.

Making the Cut

Picking the right customers has probably helped Ecco Footwear of Salem, Massachusetts, go from nothing to $35 million in sales in four years and be named footwear manufacturer of the year by the National Footwear Manufacturers Association. But don't look for them in every store.

Founder Paul Grimble picks retailers based on their ability to meet his stringent requirements:

- The shoes must be sold at full price;
- Stores must be willing to allow Ecco personnel to train and educate retail salespeople; and
- Retailers must agree to take the time and provide the expertise to service customers.

So far, those making the cut include Barney's, Bloomingdale's, Nordstrom, Overland Trading Company, and The Tannery. A pretty good team, huh?

Source: *New Hampshire Business Review,* Vol. 17, 2/3/95.

On the other hand, if we're already a regular and/or a good customer and we're giving you a bunch of our money, you certainly *should* be aware of who we are and make a point to get to know us better. And, if you want us to be any more than a place where we do our convenience shopping in the future, you'd better act as if you care about our business. As grocer-retailer-showman Stu Leonard taught us years ago, even if we're "only" a $100-a-week customer at your supermarket, keep in mind that $100 a week over a decade adds up to more than $50,000 worth of groceries—and that's if you can't convince us to up our ante by offering us prescription drugs, eyeglasses, and tires.

Oh, yes, one other thing: If we've never bounced a check at your market in the 15 years we've been writing them, no need to take down our driver's license and credit-card number each and every time we shop, especially if you recognize us. And, please, don't think you're impressing us when you read our name back to us (frequently mispronouncing it) from the face of the check.

3. LISTEN TO US, BUT DON'T BELIEVE ALL YOU HEAR —THE CUSTOMER IS NOT ALWAYS RIGHT

Contrary to what the old saw says, the customer is *not* always right. And who should know better than us? Though it's sometimes hard for us to admit, we know we've been occasionally wrong—or at least unreasonable—about the service we've demanded. More often, we've been wrong about what we thought we'd want from a company in the future. The fact is, we're frequently not even sure what we'll want next month, not to mention next year or five years from now.

Help Me Discover the Future

"If we knew what we needed, this meeting would have been over two days ago! What we want is for you to understand our business and help us discover our future and then be there before we need you to help us make it a reality."

—A chemical company executive's response to a supplier's inquiry about the coming year's requirements

It wasn't all that long ago that we were sure we'd never need a fax machine or use an overnight delivery service. As it turns out, we were wrong on both counts. For years, we paid Mail Boxes Etc. a buck and more a page to fax our stuff. Now our home fax machine gets almost as much action as our microwave (which we also thought was a ridiculous indulgence 15 years ago). As for overnight delivery, at first, we scoffed at it as needless. Now, our local UPS and Federal Express pickup and delivery people have just about become members of the family. The Internet and e-mail systems we laughed at just a year ago have become a part of our everyday work life. Truth be told, most of us continue to be amazed at how things that seem Space Age today turn into something we can't live without tomorrow.

Take computers. Many, if not most, of us were wrong about them, too. We thought we'd never end up using them in our private lives. Now, more than one-third of all American homes have at least one. And, just today, we read about a "smart" card with a built-in microchip that makes it possible to use the card in dozens of different ways. In San Francisco, you can slide the card into parking meters instead of plugging in quarters. It's even possible for the smartest of these cards to hold an entire medical history. Imagine how useful that could be to a paramedic or an emergency room doctor in the event we are ever in an accident.

No, we're not always right—especially about the future, about which, we'll humbly admit, we know very little. But please don't misunderstand. It's not that we don't *think* about the future or what we might want from companies as we enter the next century. It's just that you know so much more about your business and what's possible than we do, and there's simply no way for us to outguess you today at how you'll be there with the value we're shopping for tomorrow.

For example, it never occurred to us to ask Levi-Strauss for custom-tailored jeans. Or ask our supermarket to give us a packet of coupons based on our shopping preferences. Or beg a ski resort for a special, superfast lift line because we ski there four times a year. Or demand that we get a free mile from an airline every time we charge a dollar on our credit card. It never occurred to us to ask but, now that we have them, woe to the company that tries to take them away.

Given how much we don't know about what we'll want in the future, our advice to you is listen to us—and listen carefully. Then draw your own conclusions and figure out what we want before we know we want or need it. The companies we're likely to be loyal to in the future are not the ones we're able to order around but those that know us and know where we're headed even before we do. The companies that will come out on top will have the products and services waiting for us when we get there.

4. IF YOU WANT INFORMATION, ASK US INTELLIGENT QUESTIONS AT CONVENIENT TIMES—NEVER ASK US THE SAME QUESTION TWICE

Truth be told, we're getting pretty bored with surveys and follow-up calls as your way of finding out information about us and our experiences with you. Moreover, we seem to get so many so frequently, and we often feel that most surveys homogenize our responses. We get the feeling that we rarely see any changes as a result of participating in your surveys.

We're also frustrated by the way most surveys are written, often asking us meaningless questions—meaningless to us, that is—like whether the gas station attendant sitting in his little locked cell in the center island was wearing a tie. Who (besides your marketing department) cares whether we have 3 to 5 power tools or 6 or more? And how about those surveys the IRS is sending out after it performs an audit, asking the lucky recipient what she or he thought about the experience? Talk about a survey we're going to be *extremely* careful in filling out! Can you imagine the comments? Maybe something like, "The audit was great! We'd recommend that every American be given the opportunity to have one just like ours!"

Surveying the Surveys

Starting this year, the Center for Innovative Leadership will give out awards for the best of the best and the worst of the worst surveys. Send yours to our attention: CFIL@dsp.com. Some recent nominees include:

NBA basketball—In the city where we live, the local NBA team handed out a survey at the game—without pencils. Instead of asking us questions we *cared* about, they asked us to rate the courteousness of the beverage stand attendant. We can't think of anything we care about less at a basketball game.

Gas station survey—Come on! "Was the attendant helpful?" "Was he in complete uniform?" Who writes these things? Will someone please wake them and tell them this isn't 1955?

Internal Revenue Service—Our friend was audited recently. He was asked to complete a survey rating its service. His response: "If I mark it high, will they do it again?"

As we see it, too many surveys are bloated and jampacked with questions that either don't apply to our experience or aren't of the slightest interest to us. Worse, it's only too apparent that most of these surveys have been designed by a bunch of statisticians. Yes, they know how to tabulate the results and perform multiple regression analyses. But do they really have the requisite human touch of a company whose employees are genuinely trying to form a relationship with us?

Ten Guidelines for Better Customer Surveys Such As:

1. Short is better than long.
2. Customer concerns should predominate (both in the design of the survey and the manner in which it is distributed).
3. Involve as many people as possible in gathering and evaluating the information.
4. Ensure that the survey allows customers to differentiate among value components.
5. Comparison data are important. How do you compare with our best competitor?
6. Design the process so that the data will not be corrupted. Make getting the information a rewarding experience.
7. Listen at the periphery. All that is interesting will happen there first.
8. Seek out the most demanding customers. Through their eyes, you can almost see the future.
9. Discount any information that is exactly what you expected.
10. Noncustomers always say more about future opportunities than existing customers—if for no other reason than there are more of them—so listen to them.

In any event, if you insist on surveying us, make sure the questions are short, concise, and to the point. Keep the survey on the short side, and don't include questions that aren't appropriate for the product or service you're providing—from *our* perspective, that is. For example, if we run into your market for a loaf of bread, anything more than a single question would probably be an inconvenience. If we're flying cross-country on your airline, we sure don't want a survey that takes from New York to San Francisco to complete. Try Pittsburgh. Better yet, how about New Jersey? And, if we're at your professional sports complex, we don't want to have to answer questions about the beverage station attendant's attitude. Instead, how about a question about the sight lines in the arena or the ticketing process? Give us sports fans something we can sink our teeth into.

As for the barrage of follow-up calls we've been getting asking us our opinion of the products and services we've recently purchased from you, we're losing patience with them as well. First off, they invariably come as we sit down to dinner. Second, they're usually prescripted, and the caller usually sounds like a robot, rarely knows the first thing about the company he or she supposedly represents, and couldn't care less about our personal relationship to the company.

It's More than a Ride...

John Zimbrick, a recently retired Honda dealer in Madison, Wisconsin, devised an interesting way to listen to his customers. When his customers dropped off their cars for service, he hired cabs to take them to their homes or offices.

It was more than a ride. During the trip, the taxi drivers interviewed the customers, asking them what they thought of Zimbrick's service. Among other things, Zimbrick discovered that there are very few things a person won't tell a cabdriver. Once every quarter, Zimbrick took the drivers out to dinner and found out what customers had talked about. For the rest of the quarter, Zimbrick would concentrate on improving all the little things customers mentioned. He didn't do it once. He did it every quarter, every year, and his sons continue the tradition.

We'll give you an example: Two or three times a year, we get a call from an outside firm asking us what we think about the daily morning newspaper. We tell the caller the paper is fine as it is but that delivery has been spotty. Sorry, says the voice on the other end, I've been instructed to talk about format and content and, if we're having delivery problems, maybe we should call the paper. *Excuse me?*

Not exactly a loyalty-inspiring response, the way we see it. If the publisher is so interested in what we think of the paper, maybe he or she could hire someone from the paper to make the call—and hear us out, whether we want to talk about politics, typefaces, or having the paper tossed away from the lawn sprinkler rather than on it.

If it's follow-up calls you're making about the service or product we've received, please have people on the line who can respond to our questions, too, not just information hunters and gatherers with questions of their own. Also, make sure the caller can—and will—improvise on the script he or she is following so that we can have the opportunity to say *what* we want *when* we want it.

That Kodak Moment...On-line

Kodak is one of a growing number of companies realizing the potential of the Internet as a tool for enhancing the value it delivers to its customers. Putting its technical experts on-line in Q & A sessions allows customers to get their questions answered and enables Kodak to learn about its customers and their perceptions of the company's products—in real time.

The Internet, as Kodak demonstrates, holds great potential for getting the customer's voice quicker and faster than any survey or focus group could. Plus, it gives internal employees—many of whom are often removed from the day-to-day experience of working with customers—the opportunity to go on-line and talk directly with customers or observe customer comments, dialogues, and on-line chats.

We know of a small business that was extremely dissatisfied with the service it had received from a telecommunications vendor and was waiting eagerly for the vendor's inevitable follow-up call as an opportunity to make its complaints known. When the call finally came, the office manager launched right in, saying, "This is the worst experience we've ever, ever had...." She was interrupted in midsentence by the vendor's customer service representative, who informed the irate office manager that they had not yet reached the "remarks" portion of the questionnaire; instead, could she please rate the services on a scale of one to five, with five being the highest rating. When the office manager pointed out that the scale was not adequate to answer one of the questions, the caller told her, "Sorry, those are your only options." "And my 'remarks'?" the office manager fumed. "Remarks come at the end. Now, on a scale of one to five, was the service you received...." Honestly, you can't make this stuff up.

Remember, if you're going to interrupt our day or our dinner with your call, make sure that the call has been designed primarily to understand us better and to build a stronger relationship with us, and *not* primarily to meet your needs or satisfy your internal agenda. For example, a friend of ours recently received a follow-up call after test-driving a new sports utility vehicle. When he asked the caller what kind of information the dealership was hoping to get out of the interview and how it might be used, he was told that it was trying to identify ineffective salespeople so that the company could fire them. The call, in other words, had almost nothing to do with our friend's impressions of the test drive or whether he liked the car—only about whether the salesperson who accompanied him on his ride was doing his job the way the dealership wanted him to.

One final caution about surveys, follow-up calls, and questions about products or services in general. Make sure the questions you ask are intelligent ones, questions that cater to our own intelligence and that we're likely to answer in ways that provide you with information you can use to improve our relationship. If we've just finished eating at your restaurant, for example, don't ask us, "How was the meal?" Because, chances are, that unless someone at our table choked to death on a bone in your "boneless" chicken breast, we'll say, "Fine," the stock, get-out-of-my-face-because-you-don't-really-want-to-know-my-opinion response. As a result, you'll learn nothing. Instead, ask us whether the steak was prepared correctly, whether the chicken was the way we liked it, or whether the service was up to our standard. And, whatever you ask, make sure you're at least as gracious in accepting our criticism as you are in accepting our praise.

5. LISTEN TO THE PEOPLE WHO LISTEN TO US

We're not exactly sure what makes the airlines such an easy target for our criticism or why they're often the first to come to mind when we're talking about the ways things should *not* be. Maybe it's because they're too ambitious, trying at once to be a travel medium, snack bar, dinner house, cinema, office, and hotel. Or maybe it's because the large majority simply haven't got the hang of this customer relationship thing yet, in spite of all the information they have about us. Anyway, not long ago, a colleague of ours on a $500, two-and-one-half-hour coach flight complained to a flight attendant that the only food served was a bag of peanuts. The attendant considered his remark and then said, "Well, you should tell them about it. You're the frequent flier." "Tell *who?*" he asked. "I'm telling you. You represent the company, don't you?" "Oh sure," she said, "but I'm just a flight attendant. They don't listen to me."

Rather than viewing the airline story above as an exception to the rule, we think it's pretty typical. Lots of times, we're convinced that the companies we deal with think of their front-line employees strictly as workers, not as information gatherers. And, as a consequence, the people who actually control the work processes never ask for the information or they give it little credence.

Well, to us, these front-line employees in your company are the people we deal with on a regular basis. They *are* the company. They're the point people who are the foundation of our relationship with you, and they're the ones to whom we can best make our point—and at precisely the time and place where it's most convenient for us to make it: during the course of our business interaction. If you don't listen to them, there's no way you'll ever know what's on our minds. And the way we see it, it has to be a lot cheaper to listen to them than to commission another survey—which is usually an expensive way of finding out what we have already told you.

6. IF YOU'RE GOING TO WALK IN OUR SHOES, THEN, WALK IN OUR SHOES

But Did You Mow Her Lawn?

The way we heard it, a manufacturing company received a letter from a dissatisfied customer. Her complaint was that when she used her lawnmower, made by the company, a rock shot out and struck her leg.

In preparation for a response to this customer's letter, tests were carried out with the type of lawnmower in question. The test results gave no indication that the event described by the customer could have happened. Engineers were satisfied with the quality of the lawnmower and were prepared to discount the customer's complaint.

Armed with tests, reports, surveys, and substantial data, the engineers were confident of their conclusions. Confident, that is, until the first question asked of them by the company president was, "Did you mow her lawn with her lawnmower?" They replied that they hadn't but that they had conclusive test results to prove that the lawnmower couldn't have thrown a stone. The only response from the president: "Mow her lawn with her lawnmower."

When the engineers actually did mow the customer's lawn with her lawnmower, they found that the customer was using a company-approved attachment. When it was affixed to the lawnmower, it *did* allow for stones to be whirled out from underneath. The customer made no mention of the attachment in her letter because she didn't consider it an attachment, but just a part of the lawnmower.

The moral of the story? Customers won't always give you complete information, but discounting what they *do* tell you is risky business.

From your perspective, a 10-minute line to check your bags might seem entirely appropriate. *You* may think it's not too long but if we have a flight in 5 minutes, it's way too long. Waiting 20 minutes for a prescription to be filled may seem to you like a reasonable period but, if *we've* got a crying child at home with a painful ear infection, it's way too long. Waiting 15 days for a mortgage approval might seem perfectly okay—particularly since that's half what it took only a few years back—but, if *we've* just made an offer on our first house and we're anxious to know whether we'll be able to close the sale, it's way too long. A half-hour might not seem like an inordinate amount of time to wait for a technician to break through the Muzak on a computer help line but, if we're working on a tight deadline, it's way too long. And, finally, waiting four to six weeks for a new driver's license may seem reasonable from the commissioner of Motor Vehicle's point of view, but if the DMV has taken away our old picture ID license and replaced it with an inconsequential slip of paper that says "Renewal," it's way too long—particularly if next week we've got to show a picture ID in order to board an airplane.

INTERIM DRIVER LICENSE

R0527999 Class C

Issued: 07-16-96 518 11/ Expires: 09-13-96

Thomas George Parker SEX: M HAIR: BRN EYES: BLU
150 Arbor Road HT: 6-01 WT: 185 DOB: 08-02-43
Menlo Park CA 94025

RSTR: 01-CORRECTIVE LENSES

THIS LICENSE IS ISSUED AS A LICENSE TO DRIVE A MOTOR VEHICLE: IT DOES NOT ESTABLIS ELIGIBILITY FOR EMPLOYMENT, VOTER REGISTRATION, OR PUBLIC BENEFITS.

518 07-16-96 11/5036

The point is, if you're going to walk in our shoes, don't just slip them on and lace them up. Walk in them. Walk to the places we have to go and at the pace we have to walk. See your company and the way it affects us from our perspective. Are you really fast enough, accommodating enough, good enough to merit our loyalty? How and how much are you willing to change to become the business that no longer deals with the average customer but, rather, with one individual customer at a time?

Choosing to Care

After filing a police report for a stolen purse, our colleague started the laborious process of calling 17 credit-card companies to report the loss of her cards. Afterward, MBNA Financial Services stood out in her mind as the only credit-card company that really cared about her situation.

Although all the other companies promptly canceled her cards and told her that replacement cards would be in the mail within four days to two weeks, the MBNA response was very different. After placing a hold on her account, the representative opened up a new account and expressed his concern about our colleague being cardless for several days.

The MBNA representative asked her if it would help if he sent her a new card overnight for delivery the next morning. Could he arrange a cash advance for her from her local bank?

You know that "it's-about-time" feeling you get when someone *finally* understands your predicament? That's exactly what our colleague felt, and she's not likely to forget that it was MBNA that came to the rescue. Now she's wondering if she really needs all those other cards.

COMPANY CHALLENGES

Do you know the following things about your customers?

- Why they buy?
- How many leave? Why they leave?
- Are customers buying more than they did last year? If not, why?
- What they think you are doing well? Poorly?
- What they consider valuable in your relationship?
- How they believe you compare to your competitors?
- On each component of value, how well are you performing?
- What are the most frequently encountered problems that upset your customers?
- How often do they complain?
- How well do you recover when they complain?
- Do you understand the needs of individual customers?
- Do you understand how the needs of customers change by geographic regions and neighborhoods?
- Is it easy for customers to communicate with you?
- Who does the customer want to contact when there is a problem?

To develop a customer information system you must:

- Listen in as many ways as possible. You have to ask customers about their perceptions of service quality and solicit their improvement ideas. This takes creativity and excellent interpersonal skills.

- Write down the answers. If each member of a 10-person work team asks two customers each day, "What one thing can we do to improve things?" then, by the end of the week, you'll have 100 pieces of customer information. By the end of the month, you'll have 400.
- Analyze the information. You don't need a degree in statistics. If you take 100 of your customers' ideas for improvement and list those five things that your customers recommend most often, you will have a good idea of what matters most to them.
- Share the information. You need to get everyone in the organization involved in discovering how customers feel.
- Act on the information or pass it on to the team accountable. Try something. Continuous improvement requires action and constant experimentation. Action will also encourage customers to share more information because they know it has an impact. It's equally important to take those suggestions that you can't implement and pass them on to the people who can.

APPLICATION: Gathering and Communicating Customer Information

Begin to design a system to gather customer information that describes how the customer feels about your service and specifically what he or she would like you to do differently. Decide:

How will the information be gathered? Interviews, surveys (phone or written), suggestion boxes, shoppers—your options are limited only by your creativity.

What questions will be asked? How many of the questions benefit the customer? (Remember, if you ask only the questions that you derive benefit from, why should the customer spend the time assisting you?)

How will the information be recorded and organized to be useful? It must be understandable.

Who should receive the information, and how will it be communicated to them?

Find out from two customers per day one thing that they would like to see you do differently. Record, analyze, and share the information.

As a result of the information you gathered, list the five most frequently mentioned areas for improvement.

2.
Make Our Experience Special: Give Us Something to Talk About

Each business experiences daily moments of truth...those brief moments that occur whenever a customer comes into contact with any aspect of the company and has an opportunity to form an impression.

—Jan Carlzon

They're simple, really, the stories people tell about the special experiences they've had with different businesses. Don't let their simplicity fool you—stories are at the heart of what we feel a good relationship is about. We tell stories about our encounters with companies where caring people delivered the unusual. Stories about being different, about doing the outrageous, stick in our mind. They usually begin with the familiar refrain "You wouldn't believe what happened to me at..." As a customer, if we've experienced a "knock-your-socks-off" positive experience, we can't wait to tell everyone who'll listen. And we'll probably tell it over and over again.

Stories are a powerful medium. They make facts (or our perception of facts) come alive. They create a context for information that makes it memorable and believable. A story is heard by each listener in a unique, individualized fashion. When a story is told credibly, people will derive their own interpretations in light of their own past experiences and perceptions of the situation being described. In effect, the listener takes possession of the story and makes it his or her own.

When someone describes a restaurant as "wonderful," we might nod our heads absently. But compare that statement with a breathless encomium by a friend describing the smells, the service, the color of wine in a glass, and the sizzle of shrimp on a hot platter. When a neighbor tells us that John's Auto Repair cares, the statement doesn't have the same impact as the details of how John himself went out of his way to save her money, make the experience convenient, and give her a complete description of the work and how he encouraged her to speak directly with the mechanic who worked on the car.

Every interaction with a customer or a potential customer contains within it an opportunity to create a story. Rest assured that we won't tell positive stories if our experiences are merely satisfactory. After all, how many people want to hear about a perfectly average experience?

Tellable Tales—When Will We Tell a Story?

If you want us to recount our experience:

- It must be dramatic: Our experience must be memorable and out of the ordinary.
- It must be consistent with our past beliefs or concern subjects with which we have no prior opinion.
- It must be important to us. We will create stories only about something that is important to us. You have to get to know us through listening, information gathering, and talking to us to find what opportunity you have to help us create a story about you.

If you want us to become raving fans or what Scott D. Cook, chairman of Intuit Software, calls "an apostle"—a customer who is eager to spread the word and help convert the uninitiated—it requires something more. If you want us to be loyal and tell others about you, we must get something that's unique, something that makes us feel special.

A Caveat: Let the Seller Beware

Just as the positive stories we tell can be your most powerful advertising, our bad experiences can be your biggest nightmare. Studies show that people repeat negative stories twice as often as they do positive ones. All of us can recall at least one instance of engaging in "My-horror-service-story-is-more-horrible-than-your-service-story." They're fun to tell. They're entertaining. And it's often the one way we can strike back.

It used to be that these stories were swapped over backyard fences, by the watercooler in the office, and in the checkout line at the supermarket. Now, your worst nightmare has come to pass—the Internet. With the click of a mouse, a disgruntled customer (or employee, for that matter) can diss you on seven continents to millions of people in a nanosecond.

You want—and you need—me as an apostle. You can't afford to have me as your dissatisfied customer.

We want trusting relationships with companies that are flexible enough to address our specific needs. We want to be wowed, not merely satisfied.

Flying People, Not Airplanes

When Jan Carlzon took the helm of SAS Airlines, he developed a simple concept: "We used to fly airplanes; now we fly people."

In other words, the product being delivered was no longer defined as getting Passenger A from Point B to Point C in a jet aircraft. Instead, the *concept* of the product was broadened to incorporate travel agents, catering services, inflight crews, and baggage handlers—even the meals in coach.

Taking off on time was no longer enough. Each day, according to Carlzon's philosophy, presents the airline with 50,000 "moments of truth"—unique, never-to-be-repeated opportunities for SAS to distinguish itself favorably from the competition.

From the booking of the ticket to check-in at the counter, through the flight, and at the baggage claim area, SAS employees looked for moments to deliver superb service, a chance to create a story about SAS in the minds of its customers. In effect, SAS began to *manage* the customer's experience instead of allowing the experience to fall to chance.

To assist you in giving us something to talk about, the following four measures may help you provide the kind of experiences that not only will make us want to talk about them but that will also serve as the foundation for a long-term relationship with you:

1. Entertain us.
2. Educate us—there's nothing we like better.
3. Keep us informed.
4. Make it special or make it cheap.

1. ENTERTAIN US

Entertainment won't make us loyal, but it will get our attention. In a crowded marketplace, you'll need to work hard at getting our attention for the chance to win our loyalty. Entertainment is but one way of opening the door.

Deceptively simple and highly important, the concept of entertainment is growing. We like to be entertained when we visit a mall, fly in a plane, drive a car, or stay at a hotel. If possible, we also like being entertained by the products and services we buy and the people we buy them from. We still love it when you put on a show—more than ever, in fact.

If anything, with the commodification of many products and services, the ability to add value through entertainment has become key in a number of businesses. For example, when we travel with our kids to Disneyland, we like staying at the Lake Buena Vista Holiday Inn because they've got a kids-only check-in, a separate kids' eating area, and a live character who'll come to your room to tuck your kids in. The people working at the desk should be working the crowd in Vegas—they're that entertaining and that good.

At Oshman's, a nationwide sporting goods chain, we can do more than just purchase a basketball—we can break it in on Oshman's court. Likewise, at Don Sherwood's Golf and Tennis World, we can groove our golf swings by driving balls into a net. Flying Southwest Airlines, we're often serenaded by—and can sing along with—the flight attendants. Our local Land Rover dealer has a television on the showroom floor continuously showing its slickly produced video of the grueling Camel Trophy competition. The Bourse (French for stock exchange), a restaurant in Paris, allows the price of its special of the day to fluctuate based on demand as customers check its performance on a wall-mounted ticker tape.

Joe Boxer gives us underwear with happy faces, contented cows, and dancing pigs. Our screensaver offers us flying toasters and goofy sound effects. At Benihana Restaurants, we can watch our chef slice, dice, rice, skewer, and sauté our dinner at tableside. Our local multiplex cinema features a movie quiz while we wait for the feature to begin, and Planet Hollywood gives us a placemat with high school pictures of famous personages to match with their names. Even traffic schools have gotten in on the act with names like "Let Us Amuse You," "Laughs Galore, You Won't Snore," and "Highway to Humor."

The P.S. Factor

These days some businesses can be so entertaining that the experience becomes an integral part of the product. Years ago, we went to a gas station to fill up the tank. As we drove up, we were handed the morning newspaper, a cup of coffee prepared the way we like, had our windshield cleaned, and were handed a brochure that read, "P.S. We also sell gas."

Other interesting examples include:

- Tokyo Delvé Sushi Bar in North Hollywood, California: To the locals, it's known simply as that rock 'n' roll sushi bar but to the outside world it's a place where patrons are known to drop their eel, climb up on the chairs, and dance to the latest rock music on a moment's notice. Once inside, the interior reminds us of a cross between Barnum and Bailey and a Vegas showroom with dancing and singing waiters. *P.S. It also sells sushi, and patrons say it's better here than in Japan.*
- The Forum Shops in Las Vegas, Nevada: The floors are polished flagstone, which approximates a Roman street, and the ceiling is a painted sky whose colors change from dusk to dawn in hourly cycles. Large

Roman statues on the fountains come alive each hour and put on a show, along with the processions and nightly gladiator battles. *P.S. It also sells clothes from some of the world's most prestigious retailers—Gucci, Louis Vuitton, and Gianni Versace, to name a few.*

- Tandy's Incredible Universe Stores located nationwide: Walking into this place is like walking into Disneyland and the development labs of Apple Computer combined—it's wild, it's fun, and it's out of this world. The 186,000-square-foot stores can have celebrity appearances, a World Bazaar filled with unusual merchandise, a design-your-own-kitchen lab, a karaoke studio, educational theaters, and a high-tech day care center called KidzView. *P.S. They also sell over 85,000 different high-tech computers, home theaters, televisions, camcorders, stereo systems, and educational games each year.*
- CafeNet Inc. in Southern California: For the technically inclined consumer, it's a hip place to have a latte or java while surfing the net or playing a favorite Web site game. *P.S. It also provides CafeNet Kiosks—a partitioned off area with telephonelike booths hooked up to the Internet for busy executives to check e-mail and download files.*
- Schipol Airport in Amsterdam has more stores than a mall and boasts a casino, nightclubs, and first-class hotels that will rent a weary traveler a room for a fraction of a day. *P.S. It also doubles as a place for planes to take off and land.*

Source: Interview with Kevin Saxe, president, after finding on the Internet at http://www.cafenet.net; information from Tandy Web Page and *Discount Merchandiser Journal,* Vol. 35, 8-1-95, p. 32; information, *Los Angeles Times* Restaurant Review: Tokyo Delvé, 9-16-94.

Still, the show doesn't stop there. There are walkways at Ford, Chrysler, and NUMMI automobile assembly plants where you can watch your car being assembled. You can visit Boeing and see the wings being bolted onto a 747. It's possible to take tours of virtually every kind of business: from Napa Valley wineries, teddy bear makers, and the Rouge et Noir cheese factory in California to the Ocean Spray cranberry processing plant and the Cape Cod Potato Chip plant in Massachusetts.

The fact is, everywhere we look, someone is trying to get our attention by "making it special" with entertainment. And living in a multi-image, multimedia, multitasking, parallel-processing world, we're more than a willing audience. Therefore, anything you can give us to make our experience with you more memorable and compelling will go a long way to winning our return visit or our repeat purchase.

A Brief on Boxers

In 1985, Nicholas Graham reinvented an industry by making the plain and simple pair of men's boxer shorts entertaining. Today, his happy face boxers and dancing hot dog boxers are the underwear of choice for thousands. Graham's best seller? A pair of shorts adorned with the words "No. No. No." Turn out the light, and the boxers glow in the dark, sporting the words "Yes. Yes. Yes."

And even as Joe Boxer grows to include everything from kids' pajamas to tablecloths and a new women's line, entertainment is at the heart of the company. Not only must the merchandise be entertaining, but also the marketing and promotion are designed to engage prospective customers in the shenanigans and plain good fun.

2. EDUCATE US—THERE'S NOTHING WE LIKE BETTER

Demographers say that, as consumers, we're significantly more educated than past generations. We've bought into the concept of lifelong learning, and that hunger for knowledge appears in every aspect of our lives. Where, in the past, we left the details to others, we crave them now. The old-fashioned notion that ignorance is bliss leaves us cold. As customers, we don't like being left in the dark.

On the contrary, the way we see it, whatever you can teach us in the context of our experience with you adds value to our relationship. By that same token, whatever education you withhold subtracts value. In other words, not only do we like you more for clueing us in, we also like you less for being miserly with what you know and keeping us from knowing.

Maybe it's just because we're avid learners. We like to know how things are made and how they work. In our business-to-business dealings, your skill in educating us as part of our relationship is often even more important than the product or services. Here, we're frequently buying your expertise as much as your products and services. And, just as frequently, not knowing as much about what we're buying as you do, we need you to educate us in making the right decision. In these instances, we definitely don't want an order taker; we want an educator.

Adding Information to Your Product

- Protect your eyes: Chances are, if you enter your local Sun Glass Hut and have a question about eye care, you'll be handed a booklet complete with 40 references, six additional sources of information, and a synopsis of the latest research on ozone depletion radiation and your eyes. Material from your local optometrist? No, compliments of Revo Sunglasses.

The company routinely prepares "White Papers" for its customers and makes them available with its products.

- Connect where? Call your travel agent with a routine "connection" question and she cites the latest on-schedule statistics for six of the largest airports in the world: Atlanta: on schedule 30% of the time, but Dallas has better odds: on schedule 70%. Salt Lake City comes in at 40% and Chicago at 52%. Suddenly that one-hour layover in Chicago looks better than the 30-minute connection through Atlanta!
- The pane truth about windows: Purchase Andersen Windows and, overnight, you could become a leading expert on pane moisture, the virtues of double-paned versus single-paned, ozone waves, and tints. You thought you were buying windows when really you bought an education.
- Help desks that really help: Call Microsoft with a question and you have a wealth of knowledge at your fingertips. Without the sophisticated "help," there would be no product.
- Overnight packages from an insurance company? USAA not only sells insurance but it provides you with information about how to care for the home, the car, or the property you just insured. If you own a business, USAA will even get you a lower corporate shipping rate for overnight packages and letters.

In these and countless other cases today, the information added quickly becomes a part of the product. How can you add information to your products and services?

We particularly want to be educated in matters concerning our health and physical well-being. We want the pharmacist to be able to take a few minutes to tell us what's going on. We want the doctor to say more than

"Hmmmm" and "Ahhhh" when she's examining us and to explain what she sees and hears as the examination progresses. In any event, we certainly don't want to wait for days, or even hours, to be educated when you've already got the answer.

3. KEEP US INFORMED

As with any relationship, the sharing of information is a vital part of the foundation. We want as much information as you can give us. As a rule, err on the side of overinforming us. Open up your company. Commit to telling the truth and don't overpromise. If your product or service cannot deliver, don't make us believe that it can. You're setting us up for frustration and, when we discover the truth, our willingness to trust you in the future will be diminished. We know this sounds like Basic Relationships 101—a not-so-blinding flash of the obvious—but, for some reason, many companies have not found the truth necessary, convenient, or desirable.

It almost reminds us of the military's handling of classified information. They give it out in carefully measured doses—on a need-to-know basis. Often, we feel that what we have been given is what companies believed we needed to know in order to hand over our money. We need to change the need-to-know practices of the past to right-to-know practices for the future.

As customers, we believe that we have the right to know how any product or service might affect us or how different conditions might affect what we have purchased. We want to know if our broker is being paid in ways that might create a conflict of interest for her and a potential loss for us. We appreciate it when the pharmacist advises us about the drugs we put into our body. We are impressed when CarMax gives us the entire history of the used car before we purchase it. We were impressed when the Body Shop began informing us about the ingredients in its products in such detail as to state the country of origin and why they were selected.

And, contrary to the opinion of some, we want to understand the life insurance policies we buy, how restrictions of the homeowners' association rules might affect us, the differences among the mutual funds you are offering, and maybe even where our tax money really goes.

A simple rule—we can never have too much information. If you say that you are going to deliver the bottled water in the morning, call us and tell us if you're going to be late. If a product is on back order, give us the real details regarding the approximate delivery date. If the nonsmoking room that we've requested for our hotel stay is unlikely to be available—tell us. Don't wait for us to show up and then inform us, that because we were not the first to check in, we lose. You know how many nonsmoking rooms you have. If the fish is fresh, then call it fresh, and not some version of "fresh frozen."

I'm from The Government and I'm Here to Help You

As consumers go, we're a sophisticated lot—more educated than our parents and with more access to information than any generation of the past. Some of that education and knowledge is coming from the most unlikely sources. Consider these examples:

SEC Chairman Arthur Levitt. His unrelenting consumer focus has reshaped the Securities and Exchange Commission and marked a significant departure from the past. Under Levitt's leadership, the SEC has held town hall meetings across the country, disseminating information to even the smallest investor. The SEC has called for more training of brokers at the major houses and requested that brokers and traders and companies write their information and prospectuses in language understandable by the average investor. And, with the launch of the EDGAR system, which contains all corporate information required to be filed by law (such as quarterly reports, new stock issues, and disclosures of liabilities), into cyberspace, any person who searches the Net has access to the information reported to the SEC within 48 hours of its receipt.

The city of Fremont, California. Here, a different kind of information can be found in Silicon Valley. City Manager Jan Perkins, the City Council, and city employees attempt to communicate with residents and businesses in very customized ways:

After the Fire Department responds to a fire, it sends a letter to every resident within a 10-mile radius of the site. The letter tells the resident the cause of the fire and how the situation was handled. It also contains valuable information about fire safety and self-inspection, along with the fire chief's phone number.

The Citizens' Police Academy, where residents can volunteer for training and education with their local police force and receive instruction in crime-prevention tactics. Their knowledge of the neighborhoods they live in brings more eyes to the streets.

City-sponsored educational forums for city employees, residents, and business executives. This year's symposium featured well-known speakers on topics such as quality, customer service strategies, and strategic partners.

And say it straight. Chances are that, if you don't say it straight, it will come out crooked. Now, we know that you might not want to tell us the truth for fear that we might not buy if we were fully informed. But if a loyal relationship is your goal, there is no alternative. We appreciate, and come back to, businesses that have steered us away from certain products or services that were not right for us. Even if this means a short-term loss in revenue for you, the long-term goodwill it creates can be worth a fortune.

And it goes without saying that we want information in terms that we can understand. Common, ordinary language—and not industry jargon—will be appreciated, even if what you are telling us must be drafted by a lawyer.

Chances are, in a world where most products and services are of high quality and look like most everyone else's, the information that you add to your products will be the best opportunity to create a story—an emotional bond with us that will live far beyond the memory of any single sale or transaction.

I Want to Talk to the Person Who Knows

"No wonder many of us are loyal to our local mechanic. I recently had the opportunity to take my new car back to the dealer to have its initial service work done. Being accustomed to talking to the owner/mechanic at the local shop, I was a little surprised to be greeted by a manager who talked with us about our needs and assured us that the work would be done.

"When I returned, I was sent to a cashier who was more than willing to take my money but, when I asked her about the work that was done, she gave me a look that let me know not to ask again. When I persisted, I was allowed to talk to the manager, who then had to talk to the mechanic to answer my question.

"I then asked the manager if I could talk to the mechanic. He told us that the dealership's owner felt that the mechanic's time was better spent turning wrenches. After all, the mechanic wasn't trained in customer service.

"If this is what it got for its service training dollars, this dealership should ask for a refund. I want to talk to the people who know what's going on—the people who work on my car."

4. MAKE IT SPECIAL OR MAKE IT CHEAP

If you can't develop opportunities for storytelling, can't educate and inform or even entertain us, then our best advice is to make your products and services cheap. In the world we're moving toward, where relationships are the currency of the future, if you can't make our experience personalized and special, about the only way you'll get our attention is if your prices and services are the low-cost leader.

Many of us own businesses. Even more of us work for someone else. So we know as well as you that leading with low price can be an expensive path to take. As customers, we've subscribed to magazines at rates that wouldn't even cover the price of the paper they're printed on, signed up with a bottled water delivery service to take advantage of a two-free-bottles deal, and traveled crosstown to take you up on large quantities of your loss leader.

Make It Special or Cheap or I'll Do It Myself

With ATMs, we never have to enter a bank. Personal finance programs such as Quicken and on-line access to stocks through e-Schwab mean that we don't need accountants or brokers. With educational programs at Home Depot, we can build our own houses and repair our broken appliances. While surfing the Internet, we can book our own airline and hotel accommodations and buy health insurance. We're called the "do-it-yourself economy" and we're growing in significant numbers. If you can't make your products and services special or cheap, then we may well opt to do it ourselves. How might this trend affect your organization, your company, your industry?

While we don't feel guilty about letting our supercheap subscription lapse, canceling the water service after we've finished our introductory bottles, and buying *only* the cases of copier paper you advertised, we *do* feel that we should warn you: When we shop for price, your opportunities for building loyalty are almost nil.

So, fine, talk *only* price, advertise *only* price, and undercut everybody. But be prepared for customers who shop only price—customers that will be here one day and gone the next. When it comes to cheap, in other words, you often reap what you sow. Also, once we know you as a bargain-basement leader, it'll take a lot to convince us to pay more than a rock-bottom price for *anything* you sell, no matter how good or upmarket it is.

Now, if you want to make it special and cheap, that's a horse of a different color.

COMPANY CHALLENGES

Choosing to Create a Story: A Discussion Worth Having

We've found that the following process evokes a dialogue that can provide a mirror for the team that shows: (a) the present level of service that it provides, and (b) how much better that service could be.

1. List the opportunities you have to make an impression on your customers. Start with the customer's first contact, and trace the customer's experience throughout the process.
2. For each of these opportunities to create a story and make the customer's experience special, list what you might do to:

   a. make the customer angry.

   b. provide the customer with an average experience.

c. create a positive story that the customer will brag about.

3. Ask members of the group which level of service they normally provide to their customers. It's been our experience that, in most cases, service providers will identify the average response as the level of service they normally provide. Often, they'll smile sheepishly when asked if they ever provide service that results in the customers becoming angry.
4. Then, ask why we don't consistently provide the kind of service that creates positive stories. In most cases, there isn't a good excuse. The number 1 reason is, "We just haven't decided to be that good yet."

3.
If Something Goes Wrong, Fix It Quickly

As odd as it may sound, your best opportunity as a company to create a loyal relationship with us as customers may come minutes after we tell you that we are having a problem with a product or service that you sold us. These are the times when you get to demonstrate your commitment. In many instances, the problem won't be of your making but, as is the case in building relationships in any part of life, finding fault is rarely conducive to building trust. What counts is empathy and commitment.

Things often go wrong in ways you haven't anticipated. Even if you have anticipated a problem, solving it usually costs you considerable time, money, and effort. In any event, both of us feel vulnerable, and the natural first tendency is to protect one's interests. But we can assure you that if you will make the choice to help, it may well become the single most significant emotional connection you can make with us.

There is *nothing* more important to us than knowing that you care. We remember talking to other customers and asking them why they were loyal to certain companies. We were surprised that so many told us stories in which the company had made a mistake and then moved at the speed of light to correct it. One person told of a time when her accounting firm made an error that a freshman accounting student would not have made. When we asked her why she would be so loyal, her answer surprised us: "They aren't incompetent. They really do a good job. When they made that mistake, they moved quickly to assure me that they would take care of the problem, and they apologized profusely for the inconvenience. I never knew, before that incident, how much they cared." And now, she says, "I wouldn't switch accounting firms, no matter what."

We almost hesitate to say this, but we believe it to be true: Companies that make mistakes and fix them quickly often get more customer loyalty than those companies that do most everything right the first time. Now, we're not saying that you can make the same mistake over and over and expect that we will be there tomorrow. And we are not suggesting that you should make mistakes on purpose just to have the opportunity to impress us. (In most situations, you make enough mistakes to have plenty of recovery opportunities.) What is true for us, however, is that, when everything goes according to plan, we are getting pretty much what you have planned to give us. But, when things go wrong, *we* are the ones with a problem. To fix it, you will have to go out of your way (and depart from your normal process) for *us.* You won't go broke taking advantage of the opportunity to show us we are special.

Our Hall of Fame

Just how important is recovery? We thought that we would put it to the test. We made a list of the companies to whom we are loyal, and then asked why. Some of our reasons surprised us:

Alana's Restaurant. "The food is good, so it's often crowded, but the two women who own the place know us and always seem to find a table for us—even when the place is full. And, if something isn't right (a rare occurrence), they make it right— no questions asked."

Parkside Cleaners. "Day in and day out, it's okay, as in just above average. But, one time, I made a mistake and forgot to pick up my dry cleaning, which I needed for a big event. I called Parkside right before closing and asked if it could stay open for an extra 15 minutes. Sure, said the manager, but wouldn't it just be easier if someone from the store dropped it off at my house? And Parkside doesn't deliver—that is, under normal circumstances."

Burlingame Art Supply. "I went in to pick up art supplies when I first moved into the neighborhood. The store was small and the selection was about average. When I went to pay, I put my hand in my pocket and became instantly embarrassed—I'd forgotten my wallet. 'It's all right,' said the nice lady at the cash register. 'Just bring it in tomorrow.' Relieved, I asked, 'Will you hold the products for me until then?' 'No way,' she said sternly, 'You must take those home today.' 'But there is over $100 worth of stuff here,' I reminded her. 'So, what's your point?' she said, adding that she trusted me. 'you're good for it, aren't you?'"

Orchard Supply Hardware. "My first few visits were anything but special. Checkout was slow. The lines were longer than I would have liked, and the people were not really friendly. But then something happened while I was fretting about lawn care. I bought a few of those machines designed to make yard work easier. Most worked, but one was nothing but trouble; and so, several months later I returned it without a receipt or box but with plenty of frustration. The salesclerk didn't question my motives; he assumed I wasn't trying to rip off the store. Then, the gardening expert led me through the store, apologizing as she helped me pick out another machine. She also explained the pros and cons of different models. In the end, I had forgotten that I was in a store that was one outlet in a large chain. To me, this was my local hardware store."

Lucky Supermarket. "Although this is a superstore, I consider it my neighborhood store. After shopping for groceries for that evening's dinner, I discovered that several of the items were not in my bags but had been left on the counter by the bagger. I phoned the grocery store in frustration, actually thinking I'd have to return to pick them up. The store manager apologized profusely and asked if he could personally deliver the items to my home. Not only did he deliver the forgotten items, he added a few surprises and topped it off with a large bouquet of flowers."

Hatcher Trade Press. "I was involved in a particularly stressful project that required me to have a four-color expensive brochure printed. In looking over the proof copy, I realized that the wrong phone number had been printed on the brochure. In fact, it was my home phone number, now prominently displayed on over 10,000 five-page brochures. The job had already been run and was awaiting delivery. My frantic call resulted in a new print run, with the correct telephone number, right away. I wasn't charged for the second print run. It was our mistake, and I certainly expected to pay for my goof. Instead, I was delightfully surprised!"

When something goes wrong, there are many ways to gain our loyalty:

1. Make it fast—we've already paid.
2. Involve everyone in the recovery effort.
3. Make your recovery distinctive.
4. Be proactive—don't wait for us to complain.
5. Keep us informed—information can be as important as fixing the problem.
6. Don't make the same mistake twice.

1. MAKE IT FAST—WE'VE ALREADY PAID

It seems like many companies do far more to get our business *before* we pay than they do to keep our business *after* we pay. It's frequently the case in recovery situations that *our* money is already in *your* bank. We now bear the burden of proof for showing that we did not get the value for which we paid, that we were not contributorily negligent, and that *you* were the cause of the problem. If this sounds as though we have chosen words with legal overtones, it's because we have. Maybe it's the influence of *The People's Court* or *L.A. Law,* but caveat emptor has gotten to us; we often feel as though we have to plead a case to get you to help, marshaling the evidence to show the judge and jury—that is, you—that we are worthy and are not trying to rip you off.

While our money is sitting in your bank account and earning interest, you are missing the opportunity to create a story. Almost 60% of all service stories we have encountered involve recovery from mistakes made in the delivery process.

While a problem goes unresolved, our relationship is laid bare, out in the open, and potentially open to reconsideration. As customers, we—along with any future business we anticipate having with you—are "in play." Even if we've been loyal to you in the past, during the time it takes you to resolve our problem, we're more apt to be receptive to being courted by your competitors. And, if something genuinely attractive comes along, we are probably more likely to give it a try than we ordinarily would be. Our first question to this potential new supplier? "What would *you* do if..."

We think that the deal in a mutually beneficial relationship should be simple: The product or service we buy should work for its intended purpose. If the product fails or the service does not bring about the intended result, we should quickly work together to make sure that the value, or as much of the value as possible, is realized. To act in a manner that sends the message, "I've got your money, and now it's *your* problem," undermines the very trust that is required to cement our relationship.

Help us quickly. We've paid. The value is due. The fact that *we* have a problem makes it *your* problem. If we have been negligent, it's all the more reason for you to help us; it will demonstrate your commitment. We will understand if we helped to create the problem, and we are certainly willing to act fairly if we need to shoulder some of the cost of the recovery.

But Won't They Give Away the Store?

One of the major obstacles to providing quick response to a customer's problem is the unwillingness of many managers to delegate their responsibility to the front-line service provider. Managers often fear that employees might "give away the store."

Nothing could be further from the truth. When service professionals are required to document their recovery decisions, they are consistently more effective than their supervisors in reducing the costs of recovery. Overcontrol of recovery decisions usually results in a slower, costlier, and less effective process.

If you're not going to trust employees who serve customers, how can they possibly build the relationships that lead to customer loyalty? The solution seems obvious: Trust them, give them the authority to solve problems, and hold them accountable for building better relationships with customers.

But arguing about blame or trying to stonewall us while we are in trouble clearly demonstrates a lack of trust. If you're slow to respond in our time of need, you'll get what you deserve—angry customers who will actively work to make sure that you don't have the chance to do it to any of their friends. Recover first and talk about it later.

2. INVOLVE EVERYONE IN THE RECOVERY EFFORT

It's simple. You can't react quickly if the employee we first talk to about our problem or unsatisfying experience can't help us or must get 33 approvals from higher-ups before taking action. The person who hears our problem should *own* the problem until it's resolved. That employee should become our ombudsman for traversing the political thickets of your business. Anyone who dares tell us, "That's not my job," should be shown the door, pronto. Then it *really* won't be that person's job.

We don't expect 100% of all problems to be resolved on the phone by the first person with whom we have contact. We know that it may be necessary to consult with another person in the company. What we *do* expect, though, is that the person who first hears our problem should ensure that the ball does not get dropped. All too often, front-line people tell us they are frustrated by their inability to help and are quick to explain that they only have the authority to say no—only a manager can say yes. The result is that you've set up a system in which customers will attempt to bypass service professionals in search of the one or two managers who can help. Doesn't recovery become more expensive and less effective this way? The way we see it, recovery from customer problems must be every employee's job.

The HP Way

"My company was a subcontractor to Hewlett-Packard. It was our largest client. In one of the jobs we processed for HP, a member of its internal staff made an error in the assembly instructions.

"We were quite surprised when the entire HP team, from the truck driver to the vice president in charge of the project, appeared in our offices the following day to express their apologies for the mistake. They also wanted to spend an hour going over the scenario to determine how the error was made and how they could prevent it in the future!

"Our team members were speechless. HP's response was more about getting the problem fixed, and taking steps to make sure it wouldn't recur, than finding fault. The company acted quickly and decisively and, by involving everyone, demonstrated its commitment to the project.

"Whenever I hear the word *recovery,* I think of that meeting."

3. MAKE YOUR RECOVERY DISTINCTIVE

We won't notice what you've done to make things better if it's no different from what others would do. Recovery is show business, pure and simple. Everything your business does when we've got a problem becomes important to us. If your competitors are sending form letters in response to customer complaints, why not go a step further and give us a call?

Stories are only created around unusual situations. Sure, a less than distinctive recovery effort can turn a dissatisfied customer into a satisfied one, but that's like settling for two runs in the ninth inning to lose by only one when you could have won the game. Dare to be outrageous. We'll remember it.

What If?

Mistakes happen. Why not plan your responses in advance? It would be unthinkable to enter a new fiscal year without a financial plan or to launch a new product without a marketing plan. Addressing service problems without a plan can be just as negligent and costly. You need constantly to ask, "What if...?" and to anticipate possible mistakes. Just thinking about what could go wrong makes you better prepared when things actually do go wrong.

- What if the customer's deposit is wrong?
- What if we're short-staffed?
- What if the customer is angry after I've apologized?
- What if the package isn't delivered on time?
- What if the computer goes down?
- What if technical problems cause a power outage?
- What if the software has a bug?
- What if the 800 toll-free number for help is always busy?

4. BE PROACTIVE—DON'T WAIT FOR US TO COMPLAIN

Don't be fooled. Not all quiet customers are happy customers. Many of us, for a variety of reasons, don't or won't complain. Our problems may be small; we don't want to cause trouble or be perceived as complainers; we don't like confrontation; at times, it just seems easier to say nothing and take our business elsewhere.

You know what's on the minds of vocal, unhappy customers. But, unfortunately, the customers who complain represent only a small percentage of the total number of dissatisfied customers. If you multiply the number of customers who complain by 10 or more, you'll begin to get an accurate picture of the magnitude of the opportunity.

Just think: If you could entice more customers to complain and could convince even 10% of those to stay, think of the revenue stream that would result.

Too many companies spend too much time trying to decrease the number of complaints they receive. Strange as it may sound, we think that, if you make it easier for us to speak our minds, you'll increase the number of customers who'll be willing to give your products and services another try.

You Need More Complaints

Less than 10% of dissatisfied customers actually take the time or make the effort to complain. If a company's recovery efforts are focused only on customers who identify themselves as unhappy, then 100% of the resources devoted to recovery are focused on only 10% of the dissatisfied customers.

An effective recovery strategy must include a plan to identify the other 90% of dissatisfied customers.

Winning us back once we go is tough and expensive work. It will usually require the next vendor to be equally as ineffective. In the bad old days of the past, that might have been the case, and you could count on our coming back. But, today, with the baseline of business responsiveness rising, there's a good chance that, if we go, you will never see us again. As in, good-bye means good-bye.

5. KEEP US INFORMED—INFORMATION CAN BE AS IMPORTANT AS FIXING THE PROBLEM

It's 9 P.M. and the electricity goes out in our house. We call the power company. What do we want? Well, until a few years ago, the company probably would have said that we want our lights back on. *Wrong!* Of course, we want electricity restored. It's sweltering without the air conditioner and we've got five pounds of sirloin in the freezer just waiting to defrost...

But, first and foremost, we want information. We want you to answer the phone and not consign us to voice mail purgatory. We want you to tell us when the lights will be back on—and we want the truth. If you don't know, then, say you don't know. We understand that the power gets knocked out from time to time. Nobody can prevent a snowstorm or a hurricane. What we *don't* understand is why you would not have seen this as inevitable and had contingency plans in place to let us know what would be happening. In the Information Age, we want information, and plenty of it.

In most cases, when we have a problem, we want to know that you are working on it. We want to know that you have our best interests at heart and that you will work expeditiously to help us. Whether it's a medical care problem or the apprehension we experience waiting for our house to close escrow or waiting for an out-of-state check to clear, talk to us. They say, "Talk is cheap," but, in our case, it can still buy a lot.

Who's on First?

"During one of the recent storms, my power went off and I called the utility. After hours of dialing, I finally got through to a live person, who told me that our power would be restored the next morning. Sounded good. But it didn't come on. So I called again. By next evening, I was assured. 'Are you sure?' I asked. 'Yes,' the computer listed our street as scheduled to be back on-line by the next afternoon. Wishful thinking. So I called again.

"This time, I was told to expect power the next day by noon. By this point, I didn't believe that the utility employee taking my call had any better information than the writing guy who writes the horoscope for the morning paper, so I asked to speak with a supervisor.

"When a supervisor finally got on the line, she gave me the same answer, read from the same screen. Makes you wonder if her job description requires her to be a louder, more authoritative echo.

"When I told her my problem, she was quick to remind me that she just worked there and to tell me that the line crews do whatever they want and that she didn't think that the information the customer service reps were given meant much. She advised me not to put too much stock in it. But she did offer (because she seemed to like me) to call the office that scheduled the crews to see what she could find out.

"When she returned my call, she told me that, in truth, the utility had not yet scheduled my street but that restoring the power would be only a 15-minute job.

"Here's the clincher: As a favor to her, one of the line people fixed my problem on his way home from work.

"Several weeks later, I discovered I wasn't alone. It seems as though 70,000 other Bay Area residents experienced the same outages and similar responses. One Bay Area legislator was so bombarded with complaints from constituents in her district that she held a series of town hall meetings where residents could vent their frustrations publicly."

6. DON'T MAKE THE SAME MISTAKE TWICE

Fix your communication process. If the crews don't talk to the customer service reps, the delivery people don't talk with the sales staff, and the techies don't talk with the bean counters, it's the customers who suffer.

We will forgive you once. Sometimes twice. But we are not going to build a relationship with a gang that can't shoot straight days, weeks, or months after it's discovered that it's got a problem. When you go to the trouble of helping us with a problem, record it so that you can make sure it doesn't happen to us or any other customer again. Good people in bad processes—that's normal. Great people in great processes—that's what we want.

1-800 We Care—Teaching Old Dogs New Tricks

That was the phone number advertised by Continental Airlines for passengers to call to communicate with the company—on any issue. At least they had a number we thought. Our primary carrier didn't have one—even for their best customers.

But isn't Continental the airline of Frank Lorenzo fame, complete with troubled service and labor strife? We asked the flight attendant if the 800 number was a good idea. We just knew that we would get a cynical reply—after all the front line doesn't lie. What we heard surprised us:

"I've been here 12 years. It used to be a war but today management really listens. Just the other day, a customer didn't like the way the napkins seem to 'shed' on his suit. He called the 800 number and they changed suppliers within a week. Heck, the 800 number even works for us. We can call and they listen to what we have to say. We're not perfect, but we have made a great step towards getting better, people are beginning to listen."

Just when we had heard enough of the Continental advertisement, the man in the seat next to us chimed in, "Yea, I used to fly another airline but I switched. I get regular calls from Continental asking me what I liked and what I didn't. They're not defensive, they listen and when something goes wrong, they seem willing to fix it. I never thought I would fly Continental as my primary airline, but..."

What was that about old dogs and new tricks?

COMPANY CHALLENGES

Strategic Recovery Self-Assessment

To be effective, a recovery strategy must be fast, distinctive, proactive, and intelligently planned, with the findings of each recovery effort analyzed for further enhancement of the delivery process. The following are questions you can ask to determine whether your strategy is doing the job it should:

- Are your recovery efforts fast and distinctive?
- Are your efforts customer-friendly?
- Do you, or could you, guarantee customer satisfaction?
- Have you empowered employees to solve customer problems quickly and efficiently?
- Do your customers have easy access to the people they may need to talk to?
- Are your recovery efforts noticeably (to the customer) different from your competitors'?
- Do you give the customer the benefit of the doubt?
- Is the customer's perception of a problem considered a significant event at every level of the organization?
- Do employees have the ability to deviate from planned actions or operating procedures when they see the need to?

Do you proactively search for potentially dissatisfied customers?

- Is your goal to identify dissatisfied customers before they complain? At every level?

- Has identifying customer problems been clearly defined as everyone's job?
- Is identifying a problem perceived to be rewarding, or is it punishing to point out discrepancies?
- Are multiple methodologies (surveys, callbacks, comment cards, employee data gathering) used to identify customer problems?
- Are toll-free lines available for customers to communicate with someone who has the ability to solve the customer's problem?
- Is access to the company user-friendly and well communicated to the customer?
- Is there a structure that requires the company to act quickly on customer information when it is received?

Are your efforts strategically planned?

- Are potential problems identified as part of process improvement efforts?
- Have plans been developed to deal quickly and efficiently with common problems?
- Are these plans (and the reasons for the choices that have been made) widely understood?
- Are recovery efforts perceived to be an essential part of the delivery process rather than an afterthought?

Is process effectiveness systematically evaluated and improved?

- Are the types and frequency of customer problems tracked systematically?
- Are the root causes of problems identified?
- Is this information used to improve delivery processes as well as organizational practices?

- Do you routinely evaluate the effectiveness of your recovery efforts from the customer's point of view? In addition, after taking action to recover from a mistake, you should ask these questions in evaluating your performance in recovery efforts:
- Did your actions provoke the desired customer effect, that is, did you create a loyal customer?
- Have you used the information learned through your error and in the recovery effort to improve your process?
- Did your actions send the symbolic message to everyone in the workforce that you are committed to serving—no, delighting—customers?

Using Mistakes to Improve Service

- Make a list of all the recovery opportunities you encounter and your responses to them.
- Why did these problems occur?
- Which of them occurs most often?
- What action did you take to fix the problems you encountered? Did it work?
- Think of a routine customer interaction. List all the things that could go wrong.
- How would you recover from each of these mistakes? Explain why you chose to recover in these ways?
- How many positive service stories did you create?

4.
Guarantee Our Satisfaction

A dissatisfied customer is one of the most expensive problems you can have.

Jan Carlzon, in a speech delivered at the 1996 *Inc.* magazine Service Conference

We appreciate that you're confident enough in your product to guarantee it. What interests us even more than your confidence, however, is *if* that guarantee truly represents your commitment to us over time. In other words, what really interests us is whether the pledge that you've made and put to paper actually means that you will be there for us now and in the years to come. And whether, implicit in this pledge, there's the promise that you care enough about us to be concerned with our future satisfaction—well beyond the upbeat, exciting, and altogether positive moments typically surrounding the closing of a big deal or the completion of a purchase.

We've all experienced it; as often as some of us have been handed the keys to a new car, it's still a great feeling to drive it for the first time—with papers signed, new registration in the glove box, mirrors adjusted—taking it through its paces. The same feeling is there when we buy a new appliance or TV. We get a lot of pleasure from stripping away the chunks of Styrofoam, pulling off the protective plastic, reading the instructions in our choice of four languages, filling out the warranty card, and switching on the appliance for the very first time.

What we're really hoping to find in your guarantee—beyond the thicket of small type—is your unwritten promise that you will do everything within your power to see to it that *we will never be dissatisfied with you.* Oh, sure, sooner or later, there's the chance that we'll become dissatisfied with *what* you sold us. It may not live up to specifications or meet our needs. It might break down or turn out to be a lemon. We may even act foolishly and become dissatisfied for totally irrational reasons.

No matter. When your guarantee transcends the traditional function of promising to repair or replace a product and becomes instead an instrument that binds us together for the long term—then we, as customers, will have precisely what we're looking for: not a piece of paper promising a free repair, but a *relationship.*

If that's also what you have in mind, then, please, when you guarantee our satisfaction, do it right. And by doing it right, we don't mean a lot of misleading promises or puffery. Besides, we're not easily fooled. Most of us today can see through the smokescreen of hyped-up, pie-in-the-sky promises. The newspaper is full of advertisements that tell us that we can give it back, send it back, throw it away, get our money back, whatever. There's only one caveat: *You're* still the one who sets all the conditions; *you're* the one who makes all the rules.

That kind of one-sided guarantee may have played reasonably well in a one-size-fits-all business environment but, in a world where companies need loyal customers as least as much as customers need loyal companies, *our* rules are eventually bound to prevail when it comes to guarantees. What are these rules? Here are five of the most important ones from our perspective:

1. We want guaranteed, unconditional, subjective satisfaction.
2. The guarantee must be easily invoked.
3. There must be meaningful and significant redress available.
4. The guarantee should be clearly communicated and easy to understand.
5. If we invoke the guarantee, listen to us and remember: Our feedback is a gift.

1. WE WANT GUARANTEED, UNCONDITIONAL, SUBJECTIVE SATISFACTION

We know we're coming on strong here but, if you're going to guarantee something, it's got to be an unconditional, subjective satisfaction guarantee. After all, if Xerox can provide an unconditional guarantee on a $400,000 piece of equipment and Cooker Restaurants can provide an unconditional guarantee on *every* meal it serves, we think that, if you truly believe in your product or service, you should be able to provide us with a similar guarantee.

The way we see it, only two things stand in your way: *Trust* in us to treat you fairly and lack of confidence that your products and services meet the specifications that you create for them. On the issue of whether we'll treat you fairly, our suggestion is this: If, for any reason, you don't trust us, then, don't do business with us. Don't have us as customers. If we look or act like people who'll take advantage of your good nature and your liberal return policies, show us the door. Don't take our money today, and you won't have to worry about us haunting you tomorrow.

Guaranteed—Period

"If you are not entirely satisfied with an item, return it to us at any time for an exchange or refund of its purchase price."

—Lands' End

On the issue of faith in your own products and services, our advice to you is not to mislead or oversell us. Aside from the rare case in which a product flat out doesn't work or when it consistently breaks down—the car that's always in the shop, the computer that mysteriously crashes, the vendor who never meets a deadline—it's rare we're so displeased with a product that we'll simply give it back and demand our money. *Unless,* that is, unless you promised that the product or service could do things or perform in ways that it can't. If you misrepresent your product or overinflate our expectations, chances are that, if you've given us an unconditional guarantee, down the line, we'll be coming back looking for you to make good on it.

The important thing to remember when considering issues of trust and exaggerated expectations: The *sale* is not the thing—the *relationship* is. If you value it, make us the judge of our own satisfaction.

Out on a Limb

"I live in the same Oregon town with a tree surgeon who guarantees 100% satisfaction. Years ago, responding to his ad in the Yellow Pages, a woman hired him to prune her trees.

"Two days after the work had been completed, however, the woman called to say that her husband hated the job. It seems that he was an invalid who loved gazing at the greenery from his second-floor bedroom window, and not until the pruning was just about complete did he realize that his view would be ruined.

The tree surgeon made good on his guarantee—if the customer wasn't satisfied, he'd do whatever it took to make her satisfied. He replanted her whole backyard with mature trees. Of course, just about everyone told him that he was crazy, spending thousands out of his own pocket because the woman didn't have the sense or foresight to know that pruning would destroy her husband's view.

"Virtually overnight, his reputation grew as one of the area's most reputable and trustworthy businesspersons. Today, 15 years later, we first heard the story from a competitor, based 500 miles away from Oregon. He wishes his company would go out on a limb and guarantee its service."

---

2. THE GUARANTEE MUST BE EASILY INVOKED

We've all seen the ads—the appliance store that promises us double the difference if we see the same model amplifier we bought advertised for less within 30 days of our purchase; the credit-card company that tells us in its television spot that, too bad, if we'd used its credit card to buy that Ming vase that just fell off our shelf and shattered, it would give us our money back; the automobile dealer who's got his own private lemon law: If your new car spends more than 45 days in the shop during its first year of service, he'll replace it with a new one, absolutely free.

These sure sound like great guarantees to us—usually, that is, until we try to invoke them. At that point, in many cases, not only do we have to come up with original packaging and original receipts but we have to bring in a copy of the ad showing the lower price, model number, and a full description of the amplifier; a professional appraisal of the vase; and complete shop records of the car in question (despite the fact that the dealership has its own).

Why Aren't Guarantees Like These More Common?

Satisfaction Guaranteed Eateries—Timothy Firnstahl, the Seattle-based restauranteur guarantees happiness or your money back. His motto: "Your enjoyment guaranteed. Always."

Parrot Cellular in the heart of Silicon Valley offers a lifetime warranty on all products. Beyond the normal manufacturer's warranty, Parrot covers products it sells not for two years or three years or even five—but for the lifetime of the product.

Delta Hotels guarantees a one-minute check-in or the room is free for the night.

Flyaway Aviation Averting Systems in Seattle, Washington, has gone to the birds! The company, known for its products designed to keep birds away from airport runways, business establishments, and homes, doesn't bill the client until the client reports that the birds are gone for good.

ScrubADub Car Wash of Boston says that, if it rains or snows within 24 hours after you've left their lot, you get a free replacement wash.

The point is, if you're going to offer a guarantee of any sort, no matter how outrageously great it may be, it's not all that good for us if we have to run the gauntlet to invoke it. Make it easy on us, and we'll appreciate your effort and think about you first the next time we buy.

3. THERE MUST BE MEANINGFUL AND SIGNIFICANT REDRESS AVAILABLE

The local cable company recently made us an offer it believed we couldn't refuse. It knew that its reputation for having service representatives show up when it said they would had been tarnished by a poor on-time record. So the cable company came up with a deal: It advertised a $10 credit toward the first bill if the service rep didn't make it to our house on a schedule that, incidentally, *the company* established.

Let's get this straight. We're supposed to take two or three hours (or a whole afternoon) off from work to wait for the cable person and, if he doesn't show up, we get a $10 credit? Wow, that's great! Even better, maybe we can qualify for a second $10 credit if he doesn't show up a second time. Or is this one of those offers not good in combination with any other offer? Oh, but wait...What happens if the cable person never shows up? If we never get cable, then we'll never get to use any of those $10 credits, right? And we'll have taken all those afternoons off for nothing.

Burn the Boats

Is the waiter at Cooker Restaurant in Columbus, Ohio, more careful and more focused when he realizes that a dissatisfied customer need not pay?

Is the worker on the Xerox loading dock more careful when he realizes that, if the customer is not completely satisfied, he or she will simply demand a new machine?

Do you think Blue Cross/Blue Shield of Massachusetts worked quickly to improve its processes once people could withhold premium payments if they were not completely satisfied with the service and care they received?

L.L. Bean employees won't sell merchandise that won't wear well or that the customer doesn't need if it's likely to cost the company money in the long run.

In addition to its effect on customers, few things can make an organization focus as quickly and as thoroughly as making the customer the final arbiter of value. When the customer's satisfaction is guaranteed, a company has two choices: To serve and get paid or fail to serve and go broke. It's a Cortés experience. The boats are burned. There can be no retreat.

Any guarantee worth its salt has to be able to bite the hand that cedes it. Translation: If we have to suffer, then, as redress, you should have to suffer at least at an equal level. That doesn't mean that, if we miss the Superbowl because our brand-new big-screen TV went on the fritz, Sony has to reschedule the game. But it does mean that, if the dealer can't repair it at our house the next day, then it probably should be prepared to drop off a replacement. If it's guaranteed—whether "it" is a TV, a car, a meal, or a flue-gas desulfurization device for our electrical power plant—if you can't make it right or fix it quickly, then replace it. Don't make us pay. Or, if we've already paid, give us our money back, no questions asked. And, please, file this away for future reference: If the $20 entree is a calamity, we don't see a free $5 dessert as the answer.

4. THE GUARANTEE SHOULD BE CLEARLY COMMUNICATED AND EASY TO UNDERSTAND

When it gets to the point where we have to read the fine print in your guarantee to know where we stand with you, it's probably already past the point of no return in our relationship. That doesn't mean we're mean-spirited or unforgiving—we know that lawyers need work, too. It simply means that if you believe in our relationship, you won't make it too difficult or complicated for us to know exactly the substance and the spirit of your guarantee, right up front, on the first (and hopefully only) page. There should be nothing to hide. On the contrary, you should be proud of your guarantee; after all, it's part of the fabric of who you really are. So communicate it, and communicate it clearly.

P.O.M.G. (Peace of Mind Guaranteed)

If you are disappointed by the quality of any service we provide, we will redo the job to your total satisfaction, without the hassle, and with top priority. Period.

AND THAT'S NOT ALL...

We are committed to the belief that we can—and should—work together in close cooperation, offering you competitively priced, dependable, on-time service—available in a way that makes it convenient and attractive to work with us.

1. We promise to keep you informed about new services we offer.
2. We promise to make it easy for you to contact us through hot line numbers.
3. We promise to work to understand your special needs and concerns and to provide solutions that exceed your expectations.
4. We promise to reply promptly to all inquiries and return all calls within 4 hours.
5. We promise to give full attention and interest to your needs—and to help you look good.
6. We promise to remember that you are the customers—and our only purpose in business is to serve our customers.

These beliefs are more than guiding principles. They are the only way we will do business.

We are a part of the Kodak team and if, at any time, you don't feel the team is meeting your needs, we want to know. Call any one of us. We offer peace of mind. Guaranteed.

It's telling, for example, that virtually every person who shops at Nordstrom knows the store's guarantee, even though rarely, if ever, has anyone seen it written down. It's so well-known and so unique in the business that it has become gospel: "If you don't like it for any reason, bring it back." While we can't tell you how many people "abuse" this guarantee, we have seen Nordstrom grow and thrive as many of its competitors foundered or went out of business.

From our perspective, a One Size Fits One guarantee can't be "in the details," because every detail that we don't understand or have trouble reading is usually a provision that further protects your interests and limits ours. If you don't trust us or value our relationship sufficiently to communicate your guarantee clearly, then we're not going to wade through subclauses only to figure that out.

5. IF WE INVOKE THE GUARANTEE, LISTEN TO US AND REMEMBER: OUR FEEDBACK IS A GIFT

Our invoking your guarantee and giving you our criticism should be more profitable and of more lasting value to you than our praise—even if it costs you money. When something doesn't work for us, it's usually an indication of where and how your processes or your products may be breaking down, not only for us but also for others. Our criticism and our invocation of your guarantee is an early warning that you can use to avert further mishaps. Besides, if our relationship is on a solid footing, invoking the guarantee doesn't signal the end of our loyalty. On the contrary, depending on your speed and graciousness in seeing that our satisfaction is met, it may well signal the beginning of an even deeper, more rewarding relationship for us both.

The Love Guarantee

Jaguar automobiles come with a guarantee that's rather unconventional. The carmaker tells its prospective customers: "If you think that love isn't a sure thing, than you haven't driven a Jaguar. We will fully refund your money, if you don't love your Jaguar."

Besides, it's a fairly common phenomenon among us customers that, up until the time we give something back, we tend to feel that you owe us something for giving you our business. After we've invoked the guarantee and you've treated us royally, however, there is frequently a change in our feelings. Sometimes, it progresses to the point where we feel that now *we* owe *you,* particularly if, during the time we owned the product, we derived a certain amount of use and satisfaction from it. Whether this is always the case or not, the foundation for our relationship becomes more complex and the bond between us that much more difficult to break.

The Extended Warranty: An Endangered Species?

Americans spend $5 billion each year on extended warranties, special insurance, and service contracts. One large electronics retailer calculates nearly half its profits ($78 million in net income) from selling extended warranty service contracts. In the future, will we need to pay extra in order to have the company bear the risk of product failure, or will a new generation of companies simply guarantee what they sell?

COMPANY CHALLENGES

Crafting Your Guarantee...

- What exactly do you wish to promise and, more importantly, why?
- Have you taken steps to make certain that each individual in your organization understands the guarantee?
- Have you empowered employees to take steps to make the guarantee good? To fix processes and systems that could result in our having to invoke the guarantee?
- Are you good enough in your day-to-day operations to make the promise? If not, what do you need to set about improving?
- What is important from your customers' point of view? (A guarantee that makes customers jump through many hoops will just infuriate the very people with whom you are trying to build a relationship.)

5.
Trust Us and We Will Trust You

There's an old Nordstrom story (told to us by a long-time employee) about a salesman who was fired by one of the owners for refusing to let customers bring back products without subjecting these people to a barrage of questions. "But they're taking advantage of you," the salesman offered in his defense. "Don't do it," the owner said. "We trust our customers."

The way we see it, the owner knew what most of us, as customers, have known, or at least suspected, for a long time: That, if you trust us, treat us well, and provide us with good value, we will eventually make you rich. That didn't mean that the owner was naive and he believed that no one would ever take advantage of him, as the salesman had argued. It's just that he saw that the benefits of being known as an establishment that trusted the honesty and integrity of its customers—*without question*—would more than outweigh any losses it might suffer at the hands of those who would abuse that trust.

Though this now legendary firing took place many years ago, we're convinced that Nordstrom's belief in the importance of trust in building a successful business is every bit as valid now as it was then. Even more. While trust might seem old-fashioned—maybe even quaint—to many of those working in today's corporations, without it there can be no One Size Fits One world. Relationships require trust. Trust is at the core of every loyal relationship.

Of course, it almost goes without saying that trust has to be reciprocated. We have to trust one another. You have to trust us to be straight and honest in all our dealings with you, while we have to trust you to treat us fairly and squarely. The thing is, to a large extent, we believe we've already fulfilled a large part of the bargain. Over the years and in instance after instance, we've *already* trusted you.

We've Been Company Policy-ed

We're frustrated with being interrogated for 10 minutes when we come to return a $19.95 hair drier that we bought only yesterday and that went on the fritz this morning—just after we took the original packaging out with the garbage. We're also frustrated with the words "company policy"... as in, *"I'm sorry, I know you're in a hurry, but I need a manager's approval. It's company policy."*

Seriously, have these words ever been uttered to assure us that the company we were doing business with was looking out for our best interests?—as in, *"Of course, we'll be glad to accommodate you. It's company policy to make sure you're satisfied?"—Ever?* Probably not.

That's because *company policy* is really code for telling us we've hit the wall. No way are we going to be able to get a refund today for that blender, even though we paid cash for it—just two days before. Instead, we learn that *"All our checks are mailed from our central office. You should be getting your refund within 20 working days."*

Twenty days! Correction, 20 working days! What kind of work are we talking about here? Brain surgery? Splitting the atom? Balancing the federal budget? The second we hear '20 days' we know that, even though "My Pleasure to Serve You" is embroidered on the clerk's store vest—smack-dab over his heart, in fact— we're not an important player in this transaction. Nope, we've been company-policy-ed.

For example, we've given you money for products before we've used them. We don't know whether the lawnmower will work when we get it home or if all the parts of the telephone are in the package. Though we've paid for your health insurance policy, we haven't the faintest idea whether you'll really cover what you say you'll cover. Nor do we know whether you'll deliver what you say you'll deliver. We've paid and done so in good faith—trusting you not to let us down.

The question we pose to you today is: Do you trust us? Do you trust us to do the right thing? Do you trust us to tell the truth? *Do you believe that, given the opportunity, we won't rip you off?* Since we're the ones who usually bear the brunt of your company policies, we have to wonder. Seen from our perspective, it's these very policies—at the core of many businesses—that tell us exactly what we don't want to hear: not that we can't or won't get what we want but that we can't be trusted.

Were They the Crown Jewels?

"My family purchased a diamond ring for me for my birthday from a large retailer. I returned the ring the next day for an exchange. I was told that I could not exchange the ring because the policy was to have a jeweler examine it to make sure that I had not changed the diamonds. No jeweler was available, and none was scheduled in the store for three days.

"I asked if I had any alternatives. The clerk told me that, unless I was an 'irate customer,' there was nothing else the store could do.

"I explained that I'd be more than happy to be irate if it would help. She assured me that it would. As an 'irate customer,' I could be escorted by four employees to another jeweler within the mall, who would verify that the diamonds were real I asked why it took four employees. A saleswoman said, 'To make sure the employees don't change the diamonds during the trek across the mall.'

"*Now* I had no trouble being irate. *Four* employees to watch me? I realized it was actually one employee to watch me and three employees to watch the employee watch me.

"At this point, I realized I'd never be irate again. At least not in that store. I'm never going back."

What, for example, are most return policies if not thinly disguised procedures to keep the customer honest? What are most complaint procedures if not ways to filter out customers whose gripes aren't legitimate? That same essential and endemic mistrust of customers is the root, too, of those carefully worded guarantees that promise us the world but deliver precious little. As we see it, the most counterproductive thing you can do in trying to build a relationship with us is to perpetuate and invoke policies to protect yourself from a small percentage of people who aren't trustworthy while ensuring that the rest of us who are trustworthy are treated like potential criminals.

The point is, on the issue of trust, we believe we've already gone more than halfway. Now it's your turn. Either trust us—or don't trust us. There's no "in between." While it may have worked well in the past, these days, partial trust simply won't cut it. *We* know we're honest and that we won't take advantage of you. It's time you knew it, too, and acted as if you did.

What it all boils down to is this: If you trust us, then loyalty has a foundation on which our relationship can grow. But, if you don't trust us, that's indication enough for us that you're not looking for the same type of enduring, long-term relationship we are, but just another one-night stand.

6.
Don't Take Us for Granted

In the early 1990s, United Airlines aired a TV commercial that must have struck a nerve because United has revived it again and again, even replaying it as part of its onboard video magazine. In the ad, a company sales force has gathered in a ragtag semicircle in what looks to be a makeshift meeting room. The scene is informal; everyone is in shirtsleeves. At the semicircle's center is a tough-as-nails, maybe heart-of-gold sales manager in the Robert Duvall mode. He's lamenting the loss of a major client. He says, "Things can't go on like this. We're losing touch with our customers." His solution: that every one of the company's customers be visited personally by one of the men and women in the room. Having said this, he extracts a bundle of airline tickets from his back pocket and starts handing them out, saving the last one for himself. His personal mission will be to try to win back the major client that the firm just lost. End of spot.

We can see the appeal of this ad to the men and women

of today's savvy business world. They know—*we* know—that the relationships you have with your customers are the currency of the future. And, if this ad is about anything more than "flying the friendly skies," it's about the importance of maintaining those relationships over time and what can happen when you don't. Still, we're convinced that United wouldn't be as wedded to this commercial if it appealed only to business travelers. No, we think that the reason for its popularity is that it speaks to *us*—and most viewers, whether in business or not—first and foremost as customers. And this ad is saying precisely what we feel: Don't take us for granted.

Moreover, we like the idea that we now have the power to get an entire sales force scrambling to let us know that our relationship is important to them, and we like the idea that we've even got Mr. Big's attention. Finally, we like the thought that a *single* lost customer, albeit a good one, is enough to set all this in motion. We like it, in other words, because it's a graphic illustration of what we've always hoped our individual dissatisfaction might prompt: a One Size Fits One response.

There's no such thing as a static relationship; it's either growing deeper or falling apart. It's unlikely that many relationships can tread water for any length of time. Yet, many companies act as if a customer, once sold, will dutifully remain in the fold indefinitely. In the United Airlines commercial, the business let the relationship become "stale"—probably until it was too late.

This refusal to be taken for granted probably plays a disproportionately large role in our business relationships compared to our personal ones. Why? Perhaps because most of our commercial relationships are relatively close to the surface, so that the second we're taken for granted, we not only know it, we're already looking for a better alternative. You've courted us, plied us with all sorts of goodies, gotten us to say yes, and had a brief fling with us. Now the honeymoon is over as you look for new and better customers. And we're left with just the vestiges of your attention and concern.

How can you avoid betraying us in this way? Here are a few suggestions:

1. No taillights.
2. Don't forget who brought you to the dance.
3. Resell to us every day.

1. NO TAILLIGHTS

We were told recently that some car salespeople refer to customers they've just sold a car to as a "taillight"—meaning that, once you see the customer's taillights as he or she drives off the lot, the purchase money is in the bank and the sales commission is assured. The work of earning the business is over.

The problems inherent in the "taillight" philosophy were depicted in the movie *Breaking Away,* in which a disgruntled customer is trying to push the rear fender of a disabled heap *back* onto the used-car lot he just bought it from, while the salesman who sold it to him pushes on the front fender to keep it on the street. You can hear both sides: *"You bought it. It's your problem." "But I didn't know..." "No! You didn't ask..."*

Needless to say, thinking of any of us as a taillight is counterproductive, to say the least. The period right after you've just sold us anything is *not* the time to stop caring. On the contrary, the way we see it, it's the time to start caring *more.*

How can taillight thinking come back to hurt you? We've already pointed out the most obvious way: that, over 10 years, a loyal, $100-a-week customer will spend over $50,000 with you. In other words, let us walk away dissatisfied and that's $100 a week, *every* week, lost to you not only for 10 years but probably forever. So, if we show up at your supermarket—a first-time customer—and ask for help and we don't get it, or do get it but receive a hefty dose of "attitude" in the process, there's clearly a whole lot more at risk than what's in our cart today. The way we see it, that stark economic fact, in and of itself, should be argument enough for you not to see us as a taillight.

The thing is, that $50,000 may well understate the value of our business potential—and radically so. Say, for example, that, over the years we do business together, you expand your offerings, as so many supermarkets are doing today. Or say that our family grows or that our daughter graduates from school and starts a household of her own in the area. If any of these things come to pass, that $100 a week could easily turn into $150, $200, or more. Plus, we might well send other customers who also spend $100 a week, resulting in a total tab of maybe $300 a week that won't end up in your cash registers, simply because you refused to help us or you failed to deliver common courtesy.

A Taco for Life...

Taco Bell figures that the lifetime value of a loyal customer is $11,000. No wonder it surveys more than 800,000 customers a year in addition to the 300,000 that Taco Bell staffers talk to through the National Taco Call Center.

Supermarkets are an easy example to cite because Stu Leonard, the Connecticut grocer extraordinaire who may well be the reincarnation of P.T. Barnum, used the example when he taught us about the lifetime value of a customer. We cannot think of a business, large or small, that cannot benefit from our loyalty. Even if we do tend to purchase the same items over time, we can and do try different products, refer new customers, and spread positive stories about those companies that have garnered our loyalty.

Add it up. The numbers can be staggering. How many people have you sent to see your dentist, lawyer, doctor, printer, grocer, or electronics distributor? Find a good real estate agent and we want to save all of our friends from having to experiment with the Yellow Pages. A car dealer who not only wants our money but is dedicated to making us happy is usually at the front of our referral books. The restaurant with good food and friendly service can quickly become a hangout for all our friends. Look at the smudged and fingered cards in your Rolodex or address book...they're the ones you're always turning to, pulling out, photocopying for your friends.

Consider as well the negative effect that is inevitable when a frustrated customer is willing to utilize all his or her influence and power to prevent people from buying your products and services. We remember the old statistic that a frustrated customer tells, on average, nine other people about a negative experience. As we enter a more electronic world, how many of our closest friends will we tell in other forums? And how about the firms that will inevitably rate companies by level of customer satisfaction? Get on the wrong side of these statistics in the future and it could be very costly indeed.

It's possible that, in the past, we might not have noticed when you treated new customers better than you treated us, but no longer. We expect more from you when we are loyal. Don't rely too heavily on the past for guidance. Major League Baseball did during the 1994 strike. The owners' and the players' conventional wisdom was that fans would always come back. Well, they were wrong. These are not conventional times.

2. DON'T FORGET WHO BROUGHT YOU TO THE DANCE

Who pays the highest price for a magazine subscription? Sadly, we all know the answer. *We* do. Your *existing* customers. We subscribe to your magazine loyally for 10 years, send in our check at renewal times, never complain, barely cost you a single marketing dollar, don't even change our address. Then, out of the blue, we get an offer to subscribe to the same magazine from a high school kid who knocks on our door, or through some club we belong to—or maybe even through someone you may have sold our name to who sold it to someone else—*except* that the subscription price quoted is half of what we've been paying. Or maybe we're a big-time cellular phone user, racking up hundreds of dollars a month in charges, only to see an ad in the business pages of the morning newspaper for 240 free minutes for any new user who will sign up that day.

The temptation, of course, is to cancel our existing subscription and subscribe anew, or close our cellular account and open another, not only for the dollars we would save but for the insult you're asking us to bear.

Club Vertical Knows Each and Every One of Us

Most ski resorts don't even know the names of the people skiing on their slopes. Instead, they simply provide "upslope capacity" to transport a mass of skiers up the mountain so they can ski down. At most, they know how many tickets they have sold, and they have some notion of where people are choosing to ski based on lift lines and the volume of skiers each chairlift can transport.

The Club Vertical program at Northstar Resort near Lake Tahoe gives each member a separate ID bracelet that has a computer chip embedded in it. At each chairlift, there is a Club Vertical gate with electronic sensors that put members at the front of the lift lines. If it's a skier's first run of the day, the computer recognizes that and debits his or her credit card for a lift ticket. No more standing in ticket lines. Throughout the day and through the ski year, the computer keeps track of every run each individual skier makes. Based on vertical feet skied, individual skiers can win "frequent skier" awards and prizes.

Additionally members get discounts for using their Club Vertical wristbands with point-of-sale scanners to pay for food, merchandise, lodging, and so on. As an additional service, Northstar offers special service on slope phones for Club Vertical members. Since the system knows at any given moment what run or chairlift a club member is on, it can display real-time messages on boards at the top of each lift. As a result, a club member can be notified to call on the Club Vertical phones. That way, the resort can, for instance, relay urgent messages to individual club members from their kids.

For two years Northstar collected data on each individual customer in Club Vertical, coming to know tens of thousands of individuals—what each person liked in food or runs and needed in the rental shop, where each one liked to stay, and even such minutiae as whether a member preferred cheeseburgers over burritos on Saturday.

Based on the preferences demonstrated by each individual over the course of a ski season, Northstar decided to push its program further along the One Size Fits One path. This fall, it offered each individual skier an individualized ski package. Instead of selling customers access to slopes through the traditional lift ticket approach, Northstar offered to "presell" them the vertical feet they had skied the year before. Club members can ski at Northstar and have their accounts debited based on the vertical feet skied. This provides a new range of options and increased flexibility for these customers. Now these valued individual customers can come up Friday afternoon to ski one or two runs without paying for a full lift ticket.

The bottom line: "Our customer satisfaction measures are up in every category. In just four years, skier visits are up 300% and profits are up 400%."

And we really mean *insult.* After all, could you make our being taken for granted any more obvious than by making us, your loyal customers, pay more for your product or services than a complete stranger does? What could be more of a slap in the face than being told that our existing and continued business with you is not as important as the business you *might* get from someone else? What could be more discouraging to us than the thought that, instead of being rewarded for being your loyal customers, we're paying the freight for those special new-customer deals and promotions?

More to the point, what kind of relationship do you think treating us like that will buy you? The fact is, we're tired of seeing your superspecial come-ons and then watching as you subject those who bite—us included—to a far less appealing next week, next month, next year.

3. RESELL TO US EVERY DAY

We can't tell you how many times we've read or heard it said that we're in the midst of a "customer revolution" and that better service is the key to differentiating businesses in the Information Age. We've also heard so many talking heads on the TV say, so many times that it sounds like a mantra, that every person associated with a company or business in any way—whether that man or woman is in sales, operations, management, manufacturing, finance, or sweeping the floors—is also a service worker.

While we don't doubt that these things are true, as customers, we see today's situation in a slightly different light. Sure, everyone in business is in the business of serving customers. And, while we're perfectly happy to have you believe that everyone in your company is in the service business, what we've really noticed is that everyone in your company is also in the *sales* business.

Service is reselling. If and when we feel that you are no longer working hard to earn our business, you will be vulnerable to losing us to another company that treats us in special ways. It's not much different than any other relationship in life. Please don't forget that, after the sale, we are not taillights.

Now, that doesn't mean we want you trying to push a product or service at us every time we call with a question. That's not what we're talking about. The reselling we're referring to here is the ongoing sale of your ability to deliver the value we're looking for. You can resell us, therefore, on your skill in anticipating our needs, your responsiveness to our questions, your desire to add value to our relationship, and your interest in our continued well-being and success as a result of it. Because, if that's what you're selling, consider us an eager customer. And, if it's something that appears to be of potential benefit to us, even if it costs money, we may well be very willing to pay.

Service Is Reselling

- Build a financial model demonstrating the value of retaining profitable customers.
- Widely share retention information.
- Create a marketing budget for existing customers.
- Ensure that everyone has a service and sales role.

Aspen Ambassadors—The Volunteer Brigade

To be successful in the future, the Aspen skiing experience must be more than polite lift operators or friendly on-mountain service providers. The people of the Aspen Ski Company understand that to build guest loyalty, the guest's total experience must be outrageous in every respect. They understand that the total involvement of Ski Company employees is not enough. From airlines and hotels to taxis and restaurants, the commitment of the entire town to provide superior value for each visitor is needed.

Given that most companies have a tough time focusing their efforts on the guest's experience, is it possible to focus the efforts of a city with the diversity of Aspen? When CEO Pat O'Donnell announced that he was asking for members of the community to volunteer to give mountain tours to guests one day per week (without pay) some thought he had lost his mind. But last year, over 100 residents signed up to delight guests with their on-slope hospitality. In year two, many of the most prominent Aspen residents signed up to be part of the team—so many in fact, that it was hard to find spots for all of the people who wanted to help. Even the non-skiing residents got into the act. These Ambassadors meet regularly, set goals, share outrageous service stories, and celebrate their accomplishments. This is a town proud of its heritage and anxious to share. Today, service to guests is fast becoming a city wide obsession. This 'total involvement' is a big step toward the realization of the Aspen Ski Company mission—providing an opportunity for each guest to experience personal renewal.

7.
Our Time Is as Important as Your Time

For you, time may be money, which is probably why you mete it out so carefully and keep it under such tight control—doors that open at precisely 9 A.M. and close at 6 P.M., phones answered at 8 A.M. and not a second earlier. For us, time is...time—and frequently more valuable and highly coveted than money. Given our hectic personal schedules, the fact that most of us belong to two-income households, and that our work consumes more hours than ever before, time has become our most precious commodity.

On the job, too, we've been overwhelmed with more responsibility and larger projects with shorter deadlines all chipping away at our time. We may work for companies that have gone through downsizing, but our workloads have been upsized, and we're still making the same amount of money—adjusted for inflation, as the expression goes.

Meanwhile, our time away from the workplace—for grocery shopping, taking care of personal errands and household chores, spending "quality time" with spouses and children, or simply *relaxing*—is growing ever more scarce. Although we still have 24 hours in a day, the available time for nonwork activities is decreasing. Even our kids are feeling the squeeze, with less time with parents, more homework, more after-school and weekend activities, and more pressure to perform and compete.

All of which is to say that, if you want to have a relationship with us, you should be aware that we don't have the time we used to have and that our lives are more hectic than they've ever been before. That's why, more and more, we want to use whatever time we do have to enhance the value of our lives—and not wait in lines, wait for deliveries, or wait to cycle through your recorded announcement.

Reinventing the "Waiting Room"

We know that waiting in line, being put on hold, and sitting in the waiting room are inevitable. If we have to trade our time for waiting, at least try to make it an enjoyable experience. How we feel about the experience will include an evaluation of how we were treated while we were waiting.

In a world where relationships are everything, we give Disney the nod over our local physician. Although we agree that the waiting times at Disney can be longer than we'd like, we give Disney credit for trying to make our experience less painful and for keeping us informed. Our doctor, on the other hand, has taken the word *waiting room* too literally. Appointment times mean the times we're supposed to show up, not the times we'll actually get to see the doctor. For some reason, the doctor seems to believe that, when we're moved from the main waiting room to an exam room (really just a smaller waiting room), we don't mind.

We like the "Queue Jockeys" (QJs) that appear on the telephone lines of some high-tech companies. They give us the estimated waiting time, stock quotes, sports scores, and the weather forecast. Please, don't give us another "1,000 strings" version of a Top 40 song, and please throw out last decade's magazines and pitch the dirty watercooler.

We know that, at times, we'll have to wait—perhaps even longer than we'd anticipated. It would be nice to know that you care.

In the One Size Fits One world, therefore, it's your individual relationships with us that have to run like clockwork. And, depending on the business you're in, *clockwork* frequently means a separate clock for each one of us. Please consider the following when taking our precious time:

1. Replace business hours with customer hours.
2. Keep the appointments you make.
3. The time it takes should be the time it takes—or close to it.

1. REPLACE BUSINESS HOURS WITH CUSTOMER HOURS

In most industries, there are two times in the day in particular when we can tell whether a company feels that our time is as important to them as it is to us: when they open their doors and when they close them—also known as *business hours,* a span of time held sacred by many companies.

For example, it's a rare business whose workers don't arrive before official business hours—as posted on the door, advertised in the Yellow Pages, and reiterated on the recorded phone message. Nevertheless, it's an even rarer company whose employees breach this self-defined starting time, despite the fact that we've been standing outside the door waiting for 10 minutes for the store to open or were on the phone at 7:59, having made the foolish mistake of calling one minute before a live person was scheduled to come on the line.

Who Says We're Not Open?

"I heard that a shoe store near me was scheduled to have an outrageous one-day sale. I stopped by the store at 8 A.M., but was disappointed to find that the store didn't open until 9:30.

"As I turned to return to my car, the manager stuck his head out the door and asked if he could help me.

'When I said I was hoping to take advantage of the sale, he replied, 'Come right on in.' I said, 'But you're not open yet.' Whereupon he replied, 'Who says we're not open? I'm here, the shoes are here, you're here. I'd say it's time to open.'

"I found myself in the position of telling the manager of the store that he wasn't open. To say I felt silly would be an understatement, but my head is full of memories of pressing my nose to a plate-glass window, watching employees talking and looking at the clock to ensure that opening time wasn't pushed forward by even a minute."

If official business hour "openings" are about the guy with the stopwatch and the keys, then store closings are like the afternoon roundup. An hour before most food outlets close, the cleanup begins, and it usually involves just about everyone at the store, no matter how long the line of customers may be. At many supermarkets, 15 minutes before closing time, the P.A. system warns us that the doors are about to close: "Please conclude your shopping and bring everything to the checkout counter." And, at precisely closing time, the answering machine is switched on, the lights in the store windows are shut off, and the kindly clerk who might have been helping us just moments before inevitably shifts into fast forward. We are in the way of the employees' quick escape.

Contrast these examples of rigid adherence to closing times with those of most restaurants and the thousands of other small businesses around the country whose proprietors realize that, in times when time is hard to find, business hours are merely recommended guidelines for the time during which business is conducted.

Why not, as our friend Rick Tate would say, run on "customer hours" instead? If you say the store will open at 9 A.M., why not routinely open at 8:45 A.M.? If you post the closing time as 8 P.M., why not really close at 8:15 P.M. (or when the last customer is finished)? Why take the chance of frustrating customers whose watches are not set to the exact time or making those who show up early cool their heels?

And while we're at it, why do most companies still schedule home deliveries or repair appointments between 9 A.M. and 5 P.M.? Surely they're aware of how many households have both spouses working—or, as has always been the case—single people usually working. And don't miss the point by saying you'll be there between 9 A.M. and noon. Asking us to stay home only half the day misses the point.

Is it really too much to ask that you serve us when we're home—on the 24-hour clock and in the 24-hour world we live in? We won't tolerate *your* clock for too much longer. We need customer hours and customer policies, not business hours and company policies.

No Rest for the Weary

We need to reset our commercial clocks. Two-thirds of us—over 75 million people—put in anything but traditional business hours. We are no longer 9 to 5ers and we're beginning to live by the mantra of "These are my hours. Why aren't they yours?"

Some companies are getting it. For example, Charles Schwab, through on-line technology, lets us trade shares even if the market is closed. L.L. Bean and Spiegel do nearly 40% of their business after "standard" business hours.

As of 1996, the United States had, for example:

319 24-hour Walgreen Drug Stores

410 24-hour Safeway Supermarkets

400 24-hour 7-11 convenience stores

291 24-hour Wal-Marts

803 24-hour Kinko's Copy Shops

In the 24-hour economy there may not be any rest for the weary, but there are still opportunities to play: Take Heartland Golf Park in Deer Park, New York, which will let you tee off until 3 A.M. (Just get there before midnight, or you may face a two- to three-hour wait.)

Even more important to us as customers, the advent of pagers, e-mail, call forwarding, and cell phones means that we can usually get a message to the people we need to communicate with...at our convenience, bringing us ever closer to the personalized One Size Fits One world. Not that giving us your home number means that we'll be calling you there regularly or that giving us your e-mail address means that we'll be barraging you with communiqués at 3 in the morning. More than likely, we'll respect both your time and your privacy. But giving us the means to get a hold of you quickly and inexpensively demonstrates that our relationship is important to you—all day long.

Marriott's 11-Second Check-In

"I walked into a Marriott Hotel in Nashville, and the bellman asked for my name before I entered the lobby. When I told him, he reached up to a board, gave me my key and the contract, picked up my bags, and told me I'd be in room 211. Total elapsed time: 11 seconds.

"If the bellman was stealing bags, this was an especially good move, because I'd never seen a lightning-fast check-in, and I was speechless. But, I must say, I liked that experience a whole lot more than shifting from foot to foot while standing in a 'normal' check-in line."

In retrospect, having the key and contract ready on arrival is something Avis, Hertz, National, and all the rental-car companies have done for more than a decade. Why did it take so long for even a single hotel chain to catch on?

In a large city, a single rental-car company could have 10,000 cars of different models, sizes, and styles on the road at any given moment. No one could track the whereabouts of every car all the time. But the rental desk can have my car ready when I show up. A hotel might have 300 rooms—that never move. It turns over its inventory once a day if it's lucky. It has a guest's credit-card number well in advance of arrival. Why can't it be ready when the guest arrives?

2. KEEP THE APPOINTMENTS YOU MAKE

Inc. magazine's Entrepreneur of the Year, Flying Colors House Painting founder Bear Barnes, notes, "Eighty percent of success is just showing up because so few do."

In all truth, we wonder why we'd even have to remind you to keep the appointments you make. It's more than common courtesy (which is itself in pretty short supply these days); it's a concrete expression that you mean what you say and say what you mean.

Still, we can't count the number of times we've shown up for an appointment with the doctor, dentist, lawyer, broker, or insurance agent only to wait an hour or more. Or we pace the floor at home, waiting for the plumber, the furnace cleaner, or the electrician to show up.

We're rarely given an explanation and sometimes, not even an apology. "Imagine if I ran my business this way," we often say to no one in particular, in the wake of our frustration and anger, waiting for a painter or tile layer.

"No Pane, No Gain"

A friend in Oregon made an appointment with an auto-glass company to have her windshield repaired. The repair crew was to come to her house.

She notes, "I left work early, drove 50 miles through a hailstorm at rush hour to get home in time, and walked in to find a message on the answering machine that the repair place was overbooked and had to reschedule. *I don't think so.*"

From our perspective, being late for an appointment or missing it entirely is your way of telling us that your time is more valuable or somehow more important than ours. This may not be the message you want to send, but it is nonetheless the one we get.

What's Good for the Goose...

"My therapist has a policy common to most mental health professionals: If you miss an hour, you still have to pay for it. One day, however, I arrived at his waiting room—after having turned down assisting with a surgery in order not to miss my session—only to find a note taped to his door saying that he was sorry but that he had been called out of town suddenly and had to cancel his entire afternoon's appointments.

"Call it hostile (and he did), but when I showed up the following week for my session, I handed him a bill for the 'hour' that I missed of my own work as a result of his unexpected absence."

3. THE TIME IT TAKES SHOULD BE THE TIME IT TAKES—OR CLOSE TO IT

We've recently read that value was being added less than 5% of the time that most products are in the delivery chain. That got us wondering: First, what was happening the other 95% of the time? and, second, if that's the case, why are we waiting? More to the point, what are we waiting *for?*

If your loan committee can approve a loan in 5 minutes and you can get our credit report on-line in 5 more, why are we waiting 2 weeks and more for a loan approval? If it takes you 45 minutes to fix a bike, why do you have to have it for 3 days? When we get our car repaired, why do we have to leave it all day? When a prescription takes 2 minutes to fill, why must we wait 25 minutes? It all takes time, we've heard you say, exasperated that we'd even ask, what with all that you have to do. It sure does take time—but mostly *ours.*

8.

The Details Are Important to Us—They Should Be to You

Coffee stains on the flip-down trays [in the airplane] mean [to the passenger] that we do our engine maintenance wrong.

—Don Burr, former CEO, People Express Airlines

Don Burr's statement isn't necessarily logical or true. And it may not be fair. From a technical standpoint, it may be 100% wrong. From our point of view, however, it may be 100% accurate. We don't understand engine maintenance, but we do understand attention to detail. And to us it makes perfect sense that, if you cannot attend to small matters, why should we trust you on the bigger ones?

In a full-page newspaper ad a few years back, one airline detailed its efforts to ensure that passengers boarded clean airplanes. The ad posed the question "If their airplanes are dirty, can you imagine how the rest of the airline is managed?"

Is it fair that a piece of trash on the floor (probably left by a passenger), a coffee stain on a tray, or a lavatory without paper towels leads us to make sweeping and somewhat illogical conclusions about customer service or airplane safety? Probably not. Still, if customers are drawing these conclusions—and we are—it would be wise to manage the trivial as distinctly nontrivial.

Our initial reaction to your business is often created by little details long before we actually experience your products or services. Remember the last time you ate in a restaurant? Did someone in your party inspect the silverware and dishes to ensure that they were clean? If they were dirty, what conclusions did you draw?

Recall the last time you called a "professional" for advice. If the person did not return your call quickly, did it affect what you thought of him/her and the organization? If the person serving you is dressed unprofessionally, how much more difficult will it be for that person to impress you?

You see, our perception *is* our reality. We see the world through lenses that are unique to us. These lenses change as our experience grows, but rarely will we judge your company and the quality of its services in the way that you do. You are the experts. You understand the rationale behind what you do. You know how difficult it is to coordinate the efforts of your team. Therefore, it is understandable that you take a more rational, logical approach to evaluating your efforts.

In contrast, we usually don't have the time or the technical expertise, so we take shortcuts. We know that you understand because, as customers yourself, we feel sure that you often do the same thing and make sweeping conclusions based on your perception of the details.

Have you ever noticed that:

- When the lines are shorter, the entire bank appears to operate more efficiently?
- When the rental car is dirty, it doesn't seem to run as well?

- When you wash your car, it seems to run more smoothly?
- When the plane is late and the crew keeps you informed, the airline personnel are appearing more competent?
- When the merchandise is neatly folded, it seems of higher quality?

Make no mistake. What might seem trivial can, and does, dramatically affect your ability to build relationships with us. As one restaurant manager told us, "It's amazing how much better our steak tastes when our restrooms are clean." Of course, managing the details more effectively certainly will not, in and of itself, assure that we will build a long-term relationship with you. However, ignoring the details can kill our relationship before it even has a chance to mature.

The secret of Disney may be that it manages the details more effectively than any other organization on the face of the Earth. In our many visits to the Magic Kingdom, we have learned that, for cast members, being "on stage," in front of the guests, means that every detail from dress to actions to the cleanliness of the park will be judged by the guests and is, therefore, important.

Beware of the Freeway Theory

"You can see a lot by observing."

—Yogi Berra

It's the freeway theory. Ask someone who lives near a busy road, "How do you stand the noise?" and the response you invariably get is, "What noise?" People get so used to hearing the noise that they no longer hear it—they tune it out.

Too often the same is true in our businesses. The coffee stains that mar the customer's perception of our products and services have been a part of our operations for so long that we no longer notice them.

We need the help of a naive eye, someone to walk through our businesses with us and point out what we don't or can't see. New employees, customers, colleagues, and suppliers can be especially helpful—but only when they're convinced that we'll take the criticism seriously.

This must become a routine part of the way we manage our businesses. Only then will we develop a process to take this information and clean up the coffee stains. It's been our experience that when we ask outsiders to help us, every member of our team takes a personal interest in ensuring that we clean up our act before we stand inspection.

9.

Employ People Who Are Ready, Willing, and Able to Serve Us

I don't believe you can expect employees to have substantially better relationships with customers than management has with employees. If there's a lot of cynicism, it's going to filter through to customers.

—Author and business consultant Gary Hamel, *BusinessWeek* on-line interview, 5-5-95.

You say you want to build a relationship with us. Yet the people we deal with in your company rarely have the authority or discretion to make the decisions or do the things necessary to build that relationship. Sure, in recent years, you've given them broader guidelines to enable them to provide better service. And they do. They'll credit our account if we dispute a $10 service charge, include a few hours of free training with our next software system purchase at our request, and forgo the freight on our next shipment if our last order arrived a day late. No doubt about it, employees have more leeway in dealing with us than ever before. Enough leeway, in fact, to be able to please us much of the time.

The kicker is that loyal relationships aren't built under "much of the time" circumstances. Nor do relationships usually thrive under service guidelines that have been designed to accommodate 60% of customers 80% of the time. Most importantly, loyal relationships can't be built by people who are underempowered, underinformed, or unmotivated to do so.

On the contrary, virtually all the loyal relationships we enjoy as customers have been built by employees who have the authority, the information, and the genuine desire to serve us under *extraordinary* circumstances, when only exceptional service and One Size Fits One treatment will do. It is this level of service—and the skilled, well-trained, and dedicated employees who make it possible—that keeps us coming back again and again, no matter how special, cheap, or stiff the competition.

What can you do to ensure that these are the experiences we'll have with your company? Essentially this: Make certain that every employee in your company is ready, willing, and able to serve us to the fullest possible extent, particularly those front-line employees whom we deal with day to day. It is here, where the rubber meets the road, that most relationships are born and flourish. It is also here, where, when things don't go well, relationships crash and burn. Translation: No matter how solid our business may look, if your accounts payable representative is rude and rigid, our relationship could be seriously at risk.

If, in your company, building a relationship is not the number 1 priority in everyone's job description, then we think it will serve you well to rewrite those descriptions. Or, better yet, get rid of them altogether and replace them with a single rule: Whatever it takes, do it! In a One Size Fits One world, each and every employee—particularly every employee who deals directly with customers—should be like a family member in a family business, with the proper tools and motivation to win us over for the long term.

Here are some of the things we'll be looking for as we make that commitment.

1. Make sure employees all have the authority to build a relationship.
2. Trust the good judgment of those you are asking us to trust.
3. Empower employees to fix your system.
4. Give people the authority to give all they want to give.
5. Have intelligent, knowledgeable, well-trained service people, or don't expect us to listen.
6. All the empowerment and information in the world is for naught if the spirit is missing.

1. MAKE SURE EMPLOYEES ALL HAVE THE AUTHORITY TO BUILD A RELATIONSHIP

From our perspective, the authority that many employees have today to provide good service is merely a way station on the road to having the authority to build a truly solid relationship. As we said, we appreciate what they *can* do. By that same token, we are frequently frustrated and turned off by what they *can't* or *won't* do. Often, for example, we're on the way to building a solid relationship with a company when suddenly we hit the wall. We need to arrange a meeting at their showroom on a Sunday or have immediate access to the funds from a second-party check that we know is as good as gold. Yet the employee who's been nurturing our relationship can't make it happen even if he or she wants to, which is frequently the case.

Things were going along swimmingly, in other words, until we came up against that point where the employee's authority to build the relationship ended and the company's more restrictive policies kicked in. When this happens, often our initial tendency is to blame the employee.

But, after further considerations, we usually end up blaming the company—and almost always rightly so—and our experience sours us on the whole relationship. So, please, if you want our business for the long term, don't cripple those who are its primary builders and keepers. Instead, give them the authority to do the right thing, whatever it may be.

Dressing Room Party...

"I went to Nordstrom for its after-Christmas sale with two of my kids in tow. I hadn't planned it that way, not really—it was just the way it turned out. So here I was in a crowded store with my three-year-old holding my left had and my two-year-old holding my right. I'm not sure how it happened—I know they were with me when I went into the dressing room to try on a pair of slacks—but, suddenly, I turned around and they were gone, both of them. I panicked and was about to run out of the dressing room half-dressed when the salesperson who was helping me assured me that things were fine. He opened another dressing room door, and there were my kids and two others, sitting around a stool with orange juice, milk, cookies, and crackers on it. Rather than have the kids running around and getting lost, the resourceful salesclerk took the initiative—and probably some cash out of the register—to set up a little party. And I'll bet you anything it wasn't covered in a manual."

One thing that's certain to enable employees to build stronger relationships with us is to allow common sense to prevail, to *make common sense a common occurrence* in your company, not an uncommon one—and to give employees sufficient authority to exercise it. For example, at one time or another, we've all gone shopping for an appliance—a toaster oven, a tape recorder—seen the exact one we wanted on display, only to be told, "We're out of that item."

Our knee-jerk response? "What do you mean, you're out of that item?" So familiar is the exchange we're about to have, we already feel beaten down. "There's one right there." We point to the display model. "Right there behind the counter. I'm looking right at it."

"Oh, that," we're told. "That's not for sale. That's only for display."

Common sense tells us that, if a customer wants something you display for sale, you should sell it to him or her, even if it's the display model. The old adage to keep in mind here is that one pleased customer is always better than lots of displeased ones. Besides, if you're out of an item, why display it at all and take the chance of raising a customer's expectations and then dashing them. Remember, we're only human and, when we see something we like, we're more disappointed if we can't have it then if we'd never seen it at all. Usually, however, in cases like the "not for sale" display item, *uncommon* sense prevails—much to the salesperson's chagrin, who'd just as soon sell you the item you're both staring at than tell you there aren't any.

Compare this all-too-typical experience with one a friend of ours recently had. As he tells it, he'd invited his in-laws over that evening for a barbecue but, when he opened his old faithful grill at 5 P.M., he found it filled with water and so badly rusted that it was unusable. Desperate, he drove to the nearby shopping center looking for a replacement, finding one on the sales floor at Williams-Sonoma, a chain of home and kitchen boutiques. The explanatory card perched on the barbecue's lid noted that some assembly was required. "How much?" he asked the clerk grimly. "How much assembly, that is?"

"Not much more than an hour, hour and a half," he was told. "Oh, no," our friend exclaimed, clearly desperate. "I'm almost out of time." Without skipping a beat, the clerk made him an offer he couldn't refuse. Indicating the display model, he said "Well, this one's already assembled. Why don't you just take it?"

As with our friend in Ohio who was allowed into a shoe store before official business hours began, our friend shopping for the barbecue was stunned. "You mean you'll really let me do that?" "Of course," the clerk said—as if, in a One Size Fits One crisis situations, the standard laws of assembly were null and void. Was this an inspired act of genius on the salesperson's part? Hardly. It was simply a case of an individual with common sense and the authority to use it in the performance of his job.

The question is, why don't we hear stories like this more often? It's our guess that it's *not* because people are devoid of common sense; it's because they're reluctant to use it on the job for fear of being punished if they do. Not long ago, for example, another friend of ours ordered some bathroom fixtures from a mail-order company for a remodeling project. The fixtures he ordered included a brass faucet sprayed with a special new white finish to match the "hot" and "cold" porcelain handles and other fixtures that were part of the set. When he asked the woman taking his order whether the company would cover his return shipping if the color of the faucet didn't match the handles as advertised, he was told, "No," that the company had a policy of paying return shipping only when a product didn't work, not for any other reason.

"You mean," our friend asked, "if it doesn't work for me, like *leak,* you'll pay, but if it doesn't work for me, like *look right,* you won't?" Exasperated, the woman taking his order said, "Yes, even though I know that doesn't make sense, that's the policy." And then she added, "Of course, it wouldn't be, if it were my business."

2. TRUST THE GOOD JUDGMENT OF THOSE YOU ARE ASKING US TO TRUST

It comes down to this: If *you* don't trust the judgment of employees to take the right and sensible course in their dealings with us, how can you expect *us* to trust them enough to build a relationship? Or, more to the point, if your representative isn't capable of expressing good judgment, why would you have him or her as the point person in our relationship?

The way we see it, the only reason you'd have someone you didn't trust play a major role in our relationship was if you felt that it wasn't important enough to put into the hands of someone more capable or trustworthy. Not only is this insulting to us, it's also demeaning to the person inside your company who we're dealing with.

For example, a colleague of ours recently called United Airlines to ask for something that she knew was to be a deviation from their rules. When she finally penetrated their Maginot Line of recorded messages, recorded music, and recorded options and reached a reservations agent, she immediately said, "Listen, I need you to consider a special request that your rules don't allow, so could I just please start by speaking with you supervisor?" Instead of being connected to the supervisor as she asked, however, the agent quizzed her about her special request. "Well," our colleague said, "that'll take me a few minutes to tell you and, since I know you won't be able to help me anyhow, let's just save us both some time and let me talk to the supervisor now." "I'm sorry," the agent finally said, "but I'm not authorized to have you speak to the supervisor unless you're irate." "Okay," our friend said, "let's just solve that problem right now. I *am* irate. But, instead of having to spend 20 frustrating minutes getting even more irate, please, please let me talk to your supervisor now." The agent finally put her through.

The Phones Are Ringing

By some estimates, there are more than 6 million people in the United States who make their living serving customers on the phone. More and more companies are finding that customers will call and make contact with the company more frequently than they will take the time to write. Investment in toll-free numbers is on the rise, and projections show is that 81% of companies now utilize them to stay in touch with customers.

Customer management is now a $1.4 billion business and is growing at 90% a year. It is difficult and challenging work because the service that must be delivered, and the depth and breadth of knowledge required of the service professional, can be staggering.

The key to success is to have knowledgeable people with good interpersonal skills and a process that allows these people to make a difference to the customer. Too often, companies set up the phone system and become so focused on productivity that they lose sight of the impression that they make on the customer.

Having the phone system in place today is not only a good idea but is quickly becoming a necessity. What you deliver to the customer when you get on the phone will determine whether making the investment in the technology was a good idea or not.

The saddest thing about this story—putting aside for the moment how demeaned and ridiculous the reservations agent must have felt, having first to be the brunt of our colleague's frustration and then having to mouth the airline's silly policy—is that, even if our colleague got exactly what she wanted from the supervisor, there's no way that the experience would make her into a more loyal customer. On the contrary, she'd look for an airline with reservations agents who could accommodate her needs on the spot.

I Have to Do It All Day Long...

"Last week I went to a department store to return a $35 pair of shoes I had bought for my six-year-old son that had the incorrect cartoon figures on them. He'd only worn them for about five minutes and only inside the house on our carpeted floor. After trying unsuccessfully for about 10 minutes to get the salesman to credit my account, I finally convinced him to call his supervisor. When she arrived, without so much as looking at me or saying a thing to the salesman, she brusquely scratched her initials on the return receipt, turned on her heel, and walked away. Trying to determine what major infraction I had committed to cause her to act the way she did, I matched her stride for stride and asked, but she still refused to look at me. She wouldn't even say a word.

"That really makes me mad,' I confided to the salesperson, more out of frustration than anger.

"Makes *you* mad?' the man exclaimed. 'You only have to put up with it one time. I have to put up with it all day, every day.'"

Sure, you worry that the people who work for you won't always use the best judgment. The question is, then, do you prevent anyone except a hard-nosed supervisor from making judgments at all? Or—and this would be our preference—do you hire the best people possible, train them to use their best judgment, and then give them the authority to do so?

The way we see it, if you're running a typically tight, old-fashioned one-size-fits-all (or almost all) ship, then it's still possible to get away with having most of your employees check their brains at the door, allowing only a scattering of supervisors to exercise "judgment" while everyone else follows orders. On the other hand, if you're trying to move to a One Size Fits One existence, then the individual employees who deal directly with us have to have the flexibility to use their judgment in providing the individualized service we're looking for.

3. EMPOWER EMPLOYEES TO FIX YOUR SYSTEM

Lots of times, we find ourselves in situations where the people in a company that we're dealing with would love to be able to help us out. The problem is, they can't. Why? Because, in all too many cases, they're on the front end of an arcane or screwed-up process that simply isn't working—not for us *or* for them. Maybe it's that their company or department is not organized to work well as a team. Or maybe the company has customer service representatives in the Eastern time zone who aren't answering the phones at times that are convenient to serve the company's West Coast customers. In either case—and in the countless other cases we've witnessed as both customers and employees—these employees have to spend a good part of their day apologizing for glitches they have little control over or making excuses for things they can't, but should be able to, do.

"Sorry, Sir, I'm Not in Customer Service—I'm in Billing"

"It was a simple request, really. I just wanted to cancel a contract for a pager. After speaking to several people and listening to the company mantra of 'Our customers come first' piped in over the phone lines while I was on hold, I finally reached someone who I thought could help me. After reciting my story for the fifth time, the employee enlightened me. She said, 'You were talking to someone in customer service, but I'm in billing.' When I asked her about the customer being number one, she repeated, 'But you are in billing now.'"

For example, we've recently ordered some merchandise from a mail-order company, only to be told that it would be four weeks before we would have it in our hands. When we asked why four weeks, we were told that manufacturing could probably have it for us in three days, but that the shipping department was really slow. Now, if for no other reason than her candor, we'd be inclined to be loyal to the woman who confided in us. Every customer appreciates the cool appraisal of an insider about what is *really* going on in a company we're about to do business with. The problem is, though, once we know, we're often less apt to want to be a part of a company with ineffective processes, no matter how much we may like the messenger.

The Front Line Never Lies (Redux)

How do we know when a process in a company isn't working? Easy. All we have to do is ask. And, chances are, we'll find someone on the front line only too eager to tell us. Why this candor? Because employees are usually as frustrated as we are with the company's processes and want to go on record that they're not the ones to blame. If we give them any opening at all, they'll tell us about unfair return policies, customer service policies that don't serve customers, marketing slogans and promises that ring hollow, and billing procedures that infuriate just about everyone.

Who are these folks? They're the receptionist who greets us in the lobby, the clerk who processes our order, the salesman who takes our return, and the operator who answers our call. They're the people we deal with day to day, the relationship builders, the front line—and they never lie.

What we don't understand is, if people inside your company know about a process that's not working—and surely it seems that they're in the perfect position to know—why aren't you recruiting them to help make it better? After all, if you've given them enough authority to work with us and sometimes even enough to build a pretty good relationship, why haven't you given them the authority to change a process that, if repaired, could improve that relationship. Are you even listening to them? Or have you made it so difficult or uncomfortable for them to confide in you that they won't even tell you what they freely tell us every day?

Earlier, we suggested that you listen to the people who listen to us. As a corollary, we also recommend that you listen again—this time to the people who work the front end of your processes and who know exactly where those processes are working and, more importantly, where they are breaking down for us—and for them.

4. GIVE PEOPLE THE AUTHORITY TO GIVE ALL THEY WANT TO GIVE

We can't tell you how many times an employee—perhaps one of yours—has thrown up his or her hands in frustration and told us, "I'd like to do more for you, but I simply can't; I'm not allowed to." And yet, as we said earlier, the very "more" that this employee might have done for us may well have made the difference between a good relationship and a great one.

Losing our loyalty, however, isn't the only potential fallout from denying your best employees the authority to exercise their skill and judgment. As we all know from our own experience in the workplace, if you're continually frustrated in your work and in your ability to do a good job because of people and policies that prevent it, your dedication as an employee wanes and your commitment weakens. It's at that point that you tend to look elsewhere for satisfaction and reward. It's at that point that you tend to look for another job.

Beyond cautioning you about losing your best employees—and as customers, we know that good employees are even harder to find and replace than good customers—we'd like to offer an observation based on our experiences with most front-line employees: Maybe, just maybe, you're underestimating the intelligence and the potential of the people who serve us—which, in and of itself, could result in a sad waste of talent. In addition, these employees' frustration and anger at being thought of strictly as "doers" rather than "thinkers," may also be contagious, infecting customers with the desire to look elsewhere for a company whose employees aren't undervalued, patronized, or hamstrung from doing their best work and building the finest possible relationships. Put it this way, as employees ourselves, we prefer working *with* companies that we'd be most comfortable working for.

5. HAVE INTELLIGENT, KNOWLEDGEABLE, WELL-TRAINED SERVICE PEOPLE, OR DON'T EXPECT US TO LISTEN

Every time I found a company that had very high levels of customer or employee loyalty—breakaway levels—I'd invest in them. And, for the last three years, my investments tripled the S & P 500.

—Fred Reicheld, Bain and Company, author of *The Loyalty Effect* (Interview, *HR Focus* magazine, 6-1-96)

Nowadays, more and more of the relationships we form are based either on the data you have *about* us or the data that you share *with* us. As a consequence, if you're hoping to grow our relationship, it's up to you to put this data to the best possible use. The key to accomplishing this, we're convinced—no matter how technologically sophisticated your company or systems are—is by giving informed and well-trained service people access to that data and the charge to use it regularly in nurturing our relationship. Quite simply, no comprehensive computer program with all its bells and whistles, no elaborate electronic help desk, no tangle of telephony, e-mail, and messaging can take the place of a skilled cadre of knowledgeable employees empowered to create and build a One Size Fits One relationship.

"Please Pass the Cups to the Middle Aisle So We May Wash Them and Reuse Them on the Next Flight..."

If you've ever flown Southwest Airlines and witnessed the crew singing the welcoming song to the tune of *The Beverly Hillbillies,* watched a crew member pop out of an overhead bin or been a part of inflight high-altitude antics, you know something special is at work at Southwest. What's the *something special?* The employees.

When we asked how they decided to add comedy to the inflight routine, we got an earful. Somewhere along the line, a few employees improvised on the job, and they discovered that the customers enjoyed the "entertainment." Now the "skill" is built into the employee recruiting, screening, and hiring processes. As one longtime employee told us, "In the early years, we had to depend on one another. When it came to caring for our customers, we were allowed to do whatever we needed to do. Management trusted us—and they still do."

Herb Kelleher, the legendary CEO of Southwest, has been quoted as saying, "The Southwest culture comes from the heart and not the head."

Thank you, Southwest, for demonstrating good judgment in refusing to stop the theater. Thank you for not building a policy manual to routinize the act. Imagine instructions on the proper way to jump from the overhead baggage compartment singing your favorite Beatles song!

For example, while a computer kiosk in your waiting room might tell us about the products you offer and their availability, only a well-trained and informed employee can tell us what works under the specific and unique conditions of our facility. Or, if we are buying custom-made Levis, despite the highly sophisticated back end of this innovative and technology-dependent special-order process, only the human being at the frond end of the process can help tell us which styles and colors look good on us and which don't.

It'll be a knowledgeable human being, too—rather than a machine—who'll know his or her way through a process and tell us where things might slow down and what we can do to speed them up; how a washing machine is engineered and what that tells us about how we should use it;

what the competition offers and how they measure up (without reciting the "company line"); and which other well-informed people in your company we should talk to if the one we're speaking with can't provide all the answers.

Having said that, now that we know how much information is available to you and how cheap it is to get, there's nothing that irritates us more than dealing with people in your company who are woefully underinformed, for whatever reason. Where once we could understand why a teller, for example, might not know that we're a good customer or an automobile salesman might not know his or her product in addition to that of the competition, today that sort of ignorance is pretty inexcusable.

Even more undermining to our ongoing relationship than our irritation with underinformed employees is our frustration and impatience with employees who haven't been trained well enough to use what information they have. We're tired of salespeople who have only a veneer of knowledge about what they're selling and precious little instruction on what to do with it—particularly with high ticket items like cars. We're tired of paying huge commissions to real estate agents who can't even navigate the simplest deal without making serious mistakes. We're tired of stockbrokers whose only "research" is what the company passes on to them.

Once again, if it's a relationship you want, in the One Size Fits One world we're moving toward, you have to spend the dollars and the time to make the men and women in your organization the most knowledgeable in their field. Only then can we feel confident in their counsel and certain that we've brought our business to the right people and the right place.

6. ALL THE EMPOWERMENT AND INFORMATION IN THE WORLD IS FOR NAUGHT IF THE SPIRIT IS MISSING

When it comes tight down to it, we like doing business with people who like doing business with us, whether we're buying a muffin to eat along with our morning coffee or working with a vendor in a business-to-business relationship. When the people we deal with demonstrate a love and interest for their work, their spirit is infectious, and the possibility of their gaining our loyalty is greatly improved. On the other hand, if the people we work with seem merely to be going through the motions, no matter how empowered they are or how much information they have for us, their lackluster attitude won't buy their company much more than the bare minimum of our attention, and little if any of our loyalty.

Did You Color Outside the Lines?

The MTV-like rap musical show being played on a large-screen television for all prospective employees to watch featured the CEO and Southwest employees "singing," dancing, and basically "living" the Southwest spirit. As Monah Hernandez of Southwest Airlines' Northern California office told us, "We are always on the lookout for people who colored outside the lines as a child." That is the beginning of Southwest's quest to bring each and every customer "positively outrageous service," or P.O.S., as it's referred to at SWA. As Hernandez explained, "At Southwest, we believe we can teach any employee a skill, yet the right attitude is an inherent quality and cannot be taught. Our efforts to hire the right attitudes start with a 'fun' theme."

It's not unusual to be asked in an interview with Southwest to explain how you used humor in an embarrassing or difficult situation. Also, don't be surprised if

Southwest customers are a part of your interview. The airline started using customers in the selection process several months ago. As Hernandez put it: "Who better to help us select our employees but our customers?"

The way we see it, business is ultimately a very human enterprise, and the way we feel about the people we're doing business with is every bit as critical to us as the final result we're seeking. If the people we do business with communicate an excitement for their work and for building our relationship, we'll go out of our way to find a reason to continue to work with them. If we don't sense that spirit, however, no matter how otherwise positive the outcome, we'll usually find a reason to move on. Spirit is like a magnet. If it's there in the companies we do business with, we'll be attracted to it—just like we are to the college football team, the actress, the politician, or any individual with spirit. On the other hand, if that spirit is lacking, there usually won't be enough in its stead to continue to draw us in.

Ladies and Gentlemen Serving Ladies and Gentlemen

Long regarded as one of the bastions of great customer service, the Ritz-Carlton chain of hotels isn't always "up front" about their success. Their secret? "Ladies and gentlemen serving ladies and gentlemen"—meaning that courtesy, respect, and commitment to service extend to the back of the house among *all* its employees as well as the front. Everyone on the Ritz-Carlton team, in no matter what job, is treated with the same dignity and concern as the customer.

Of course, you don't have to be the Ritz to make treating employees with respect work to your benefit and that of your customers. Just as successful are companies as diverse as the Walt Disney Corporation, Hewlett-Packard, Price/Costco, and, of course, Nordstrom.

How do you go about creating this spirit? Money—frequently the first thing that comes to mind when thinking about ways to motivate people—is a factor but, as far as we're concerned, it's not *the* factor. Not even close.

By way of illustration, we recently visited a number of outlets of a national restaurant chain where the pay was the same from restaurant to restaurant. Nevertheless, in some of the restaurants, the service was impeccable and the spirit bracing while, in others, the service was slapdash and the only spirit to be had was in the drinks at the bar.

Conduct Your Own Study

We are willing to bet you our business that the branches, outlets, or stores within your company that have the highest levels of customer loyalty also have the highest levels of employee loyalty. Pick the best performers—consider revenue, profit, customer satisfaction, and customer loyalty, and then compare the employee satisfaction and loyalty of these stores with those that are performing at a lower level.

What made the difference? Clearly not money in this case, since money was a constant. Rather, based on our own experience on the job, we have to assume that spirit originated where it almost always does—with leadership—the way each restaurant manager set expectations for employees, treated them as a team *and* as individuals, and gave them the authority and the training necessary to get the job done. That, and encouraging everyone on the staff to choose to serve rather than serve to get.

We've also seen employees in the same industry who were paid twice what others were and who were nevertheless no better and no more interested in forming a relationship with customers than their lower-paid brethren. In this case, as in most cases we've experienced as customers and employees, money was *not* a motivator—not a motivator, that is, to increased desire or greater caring, the standards we apply to a truly exceptional employee in a One Size Fits One environment.

By that same token, our own experience also tells us that not having enough money to live can be a serious *de*motivator. A golfing buddy of ours, for example, recently told us that he was getting incredible deals on golf equipment—considerably better than the 5% over cost we'd managed to find at our local cost-cutting golf and tennis shop. "The thing is," he confided, "it's got to be a cash deal—as in, you give the guy at the golf shop $300 in cash and he'll give you a $1,000 pair of irons"—meaning, of course, that the employees were stealing the merchandise and selling it under the table.

We wondered why employees would resort to stealing from their boss and asked our pal if he knew how much they were paid. "Minimum wage plus commissions," he said. "Commissions on everything?" we asked. "Nope," we were told, "only on slow-moving items. The policy is, the second an item takes off, the boss takes it off the commission list."

We do not condone stealing. However, if you don't give people the opportunity to earn a living, they'll find a way, and frequently at a far greater expense to you than if you had created a more equitable—in terms of dollars and cents—working environment for them.

They Can't Afford to Stay

"I was called on to work with a self-service giant, whose stated desire it was to be the Disney or the Nordstrom of the self-service industry. To accomplish this, the company sent the whole management team to be trained at Disney, got Disney name tags, Disney this and Disney that. They Disneyized the entire company...all they forgot was the Disney spirit.

"A few months later, I returned to the headquarters and, unbeknownst to the management, stopped in at one of the company's downtown Los Angeles stores to check things out. In my brief time there, I was quoted incorrect prices, virtually ignored by three employees, and insulted by the one of the people working the registers; I also saw that the store was incorrectly stocked.

"Later that day, I asked the store manager, 'How's it going?' "Okay,' he said. Suspecting what might be the problem, I asked, 'Well, what kind of turnover are you experiencing?' 'About 85% in the last nine months,' he told me. Eighty-five %! 'You know,' I said, 'it's going to be hard to be the Nordstrom of the self-service industry with employees who don't know the business and whom you can't afford to train because they probably won't be there long enough to learn anything.'

"Finally, mindful that we were in the middle of a relatively expensive metropolitan area, I asked him how well he paid his employees. 'Well, we only pay $4.60 per hour, and the best employees get jobs with other companies. It's a real problem. No wonder, I thought. No way you can get loyal employees, much less loyal customers, if you're paying less than what people can live on. It's the old saying in action: You get what you pay for."

On the other hand, a bottled-water consortium takes a far more enlightened tack. They figure that you have to pay people well enough to live comfortably. And, as long as the company keeps making money, it's a win-win situation.

The way we see it, it all comes down to this: If you seriously underpay people, their tendency is to be uninterested in us or a relationship of any sort. Or worse, if their low pay is augmented by other incentives that encourage them to sell us things that we don't want or need, they'll usually do whatever it takes to make the sale...and almost always at the price of our trust and our future with your company.

One final thought: Asking us to form a relationship with someone in your company who doesn't like relationships—or is unmotivated, for any reason, to form them—is crazy. If you want people—as in *customers*—to work with *you*, put people on your payroll whose first priority is to work with us...and to do so knowledgeably and gladly.

The Service Business Is the People Business

If you don't enjoy working with people, if you don't like serving people, if you don't like hustling and customizing experiences for demanding customers, find a new line of work.

Simply put, the service business is the people business.

There are no exceptions or excuses.

10.

We Care Whether You're a Responsible Corporate Citizen

Everyone wants to jump on the corporate responsibility bandwagon. It would be great if everybody meant it, but too many want to raise the corporate responsibility banner to boost morale, improve productivity, and get the media to quit calling them job-killers. I tell them it's all about values and caring and treating people like you'd want to be treated. Their eyes glaze over. They want the quick and easy answer.

Elizabeth Sartain, Vice President, People Department, Southwest Airlines

Information cuts both ways. At the same time that you're learning more about us, we're learning more about you. We're reading about you in the papers—and not always in the business pages—and hearing about you on TV and radio—in business reports and on consumer hotlines. We know when you've laid off employees, when your tanker runs aground, or when your employees are suing you. We also know about your good works: your support of local charities; your enlightened policies to protect natural resources; and the high quality of your own working environment. In other words, we know, as Santa does, whether you've been naughty or nice. And it makes a difference...to us.

Give Kids the World

Give Kids the World is one of our favorite nonprofit organizations. Located in Orlando, Florida, it fulfills the dreams of terminally ill children whose last wish is to visit with Mickey Mouse. Through the passion of its founder, Henry Landwirth, and the generous donations of many corporate sponsors, each year more than 4,000 families get to visit Disney World, Universal Studios, Sea World, the Hard Rock Cafe, and other Orlando attractions.

Just how much do organizations really donate? Take, for example, Budget Rent-a-Car. It's been providing cars for families since the first visitors to the Kids' Village. What started as four cars per week, grew last year to cars for 4,000 families.

Perkins Restaurants serve more than 110,000 free meals a year to the kids and their families. Companies like Disney, Universal Studios, and the Hard Rock Cafe not only provide access to their attractions, but many employees from these and other Orlando companies contributed their labor to help build the Village.

And that's just the tip of the iceberg. Proctor & Gamble, Sprint, American Airlines, and Kmart, among others, also make generous contributions without asking for anything in return. To witness firsthand the joy that Give Kids the World brings to sick children and their families in their time of need, and to understand the commitment of its corporate partners, is to experience the very best of the American spirit.

Although the corporate partners do not and probably should not advertise their involvement, we are beginning to learn who is willing to help those in our communities who are in need. The more we as consumers know about companies that care and reach out, the more we can reward them for their efforts.

Give Kids the World and its sponsors deserve our heartfelt thanks.

After all, in a One Size Fits One world, we're not just doing business with you; we're carrying on a relationship. As such, we don't want to be ashamed of how you act, what you say, and what you do. We want you to act humanely, to live by commendable values, to treat others fairly and well (customers and employees alike), to contribute to society, and to protect the environment. We want to feel good about who you are and the way you do business—just as you say you want it.

According to an *Inc.* magazine fax poll, "Customers are now more inclined than ever before to make a purchasing decision based on how they feel about a particular company" in terms of social responsibility.

Just as we do when you refuse to take the time to know who we are, take us for granted, or treat us as if we didn't matter, we'll take our business elsewhere if we don't like what you do or what you stand for. Not long ago, for example, we read about a customer who switched long-distance carriers after the company he had been using downsized, laying off 40,000 workers. Asked to comment, he said, "I can't see giving my money to a company which finds people that expendable. They made my decision easier."

What Values?

Of the 67% of employees who consider a "code of values" either very or somewhat prevalent among business, only 7% think that companies actually live by those values.

Was this customer the Lone Ranger? Or was he just one of a growing number of customers who are downsizing their financial support of companies that act in ways inconsistent with those customers' ideals and sense of fair play? We'd put our money on the latter.

> We're in the steel business. We have not laid off an employee or shut down a plant in the past 23 years. And we have a saying: Share the gain and share the pain. When times get slow we go to three-and-a-half or four-day workweeks. Nobody is worried about their jobs, nobody is getting laid off. We all take a pay cut—all of us. So the boat either rises and everyone goes up together or, when the boat sinks a little, everybody sinks together. And you sure develop a heck of a lot of employee loyalty and trust. You have to be willed a job in one of our established plants today.
>
> **—John Correnti, Nucor Corporation**

Now, if all this sounds like soft-headed claptrap to you, beware—it isn't. Today, customers of *all* stripes and political persuasions are joining the fray to bring greater social responsibility and the Golden Rule into the corporate world. What used to be fringe groups with special interests have become mainstream—liberals and conservatives, Republicans and Democrats. Over the last decade, local and regional boycotts have mushroomed. Many turned into major national and international movements, such as those to save dolphins from the nets of tuna fishermen and eliminate animal testing in the manufacture of cosmetics. Before the dismantling of apartheid, pressure on organizations to divest themselves of stock in companies doing business in South Africa was a potent force.

The Handwriting on the Wall

78% of us are currently avoiding or refusing to buy from certain businesses because of negative perceptions about them. Of that number, 48% said that unethical or unlawful business practices play a significant role in determining our decisions.

75% of us say that our purchasing decisions are influenced by a company's reputation with respect to the environment, and 8 in 10 of us say that we'll pay more for products that are environmentally friendly. From 1985 to 1990, "green" product introductions increased at an average annual rate of 100%, and sales of green products have reached $8.8 billion to keep up with our demands.

47% of us indicate that we would be much more likely to buy from a company that is socially responsible and a good corporate citizen if quality, service, and price were equal to that of the competitors.

A growing group—16% of us—actively seek information about a company's business practices before purchasing.

$650 billion, or $1 out of every $10 we invest, goes to companies that meet some specific social criteria (up from $40 billion a decade ago).

A Yankelovich Survey found that, by a 3 to 1 margin, Americans believe businesses have a greater obligation to their workers than to their bottom line.

One look at today's supermarket shelves, laden with products sporting "environmentally friendly" packaging, should be enough to convince anyone that the movement to more responsible corporate citizenry is catching on in a very big way.

Clearly, the consumer's voice is being heard and, just as clearly, many of the nation's leading corporations are paying heed to what we're saying. Even golf courses have caught good citizen fever, changing the fertilizers and insecticides they use and planting hardier grasses that cut down on the need for frequent energy- and water-consuming irrigation.

No Sweat

A number of American clothing and footwear manufacturers were caught unawares this year when human rights groups, along with the U.S. Department of Labor, identified their goods as being made with child labor.

The "but-I-didn't-know" response was seen by most as irrelevant. Clearly, today's consumers feel that companies bear the responsibility for knowing the conditions in which their goods are produced.

Embarrassed and apologetic, a number of these companies joined with other concerned manufacturers to promote a "No Sweat" label for clothing and shoes made by workers in nonexploitative environments. Secretary of Labor Robert Reich announced that the government would routinely publish a list of retailers and manufacturers whose products meet human rights criteria.

Information, coupled with our willingness to withhold our money, can change entire industries almost overnight.

Conscience and altruism notwithstanding, at the heart of this growing response to our call for greater corporate responsibility is *money*—the money we're withholding, both as customers and shareholders, from companies whose business practices we find offensive. But that's fine with us. The way we see it, if it takes the threat of a diminishing bottom line to make some of our wealthiest and most influential corporate citizens sit up and take notice, then, so be it.

A $31 Billion Good Citizen

Shareholder return is important. No company can survive for long if it does not provide an adequate return to its investors. And, while some corporate executives think that profitability is the sole purpose of a business enterprise, most people feel differently. David Packard, cofounder of Hewlett-Packard, recounted his experience at a business conference whose main topic was corporate responsibility. He remembered how the general consensus among his peers was that the only duty of a corporation is to its shareholders. "I think you're absolutely wrong," said Packard. "Management has a responsibility to its employees, it has a responsibility to its customers, and it has a responsibility to the community at large." He noted later that the other attendees "almost ran me out of the room."

It's not that Packard's commentary is so unusual. What is unusual, however, was how David Packard lived those values. Today, HP is an exemplary corporate citizen with over $30 billion in annual sales. Its commitment to philanthropy, education, and community enrichment has endured through good times and bad. And Packard was a noted philanthropist in his own right.

As consumers continue to find their collective voice and demand a closer link between business operations and good citizenship, we're likely to see more leaders like David Packard on the covers of business magazines and fewer corporate executives who satisfy shareholders at the expense of their employees and the communities in which they do business.

Businesses should also know that our interest in their good citizenship is not just a passing fancy; we'll continue to monitor their actions carefully. In fact, the way we see it, a company's demonstrated commitment to the larger world of which it is part could well be its trump card for entering the 21st century in a leadership position.

Three More Corporate Good Citizens

- Working Assets, named one of *Inc.* magazine's fastest-growing privately held companies (for four years in a row) sells local and long-distance telephone services and offers its customers a credit card. The real story may be the firm's dedication to "passive philanthropy." With 230,000 residential customers, revenues growing from $2 million to $100 million in a span of 24 months, and 100,000 credit-card customers, the firm donates 1% of its customers' bills to 36 nonprofit groups. In 1995 it donated a total of $2 million to such groups. In addition to philanthropy, Working Assets offers its customers free monthly calls to "exercise democracy" by phoning elected officials on issues as diverse as protection of the environment and public education. Customers took the firm up on its offer by making nearly a million calls.
- Yvon Chouinard, founder and chief executive officer of Patagonia attempts to bring to life the stated goal of the company: *Make the best-quality product and cause no unnecessary harm."* Chouinard stated that he didn't believe it was possible to make a great-quality product without having a work environment that made it possible. Patagonia's approach is to link quality product, quality customer service, and quality of workplace and life. If any one piece is missing, there's "a good chance you'll miss it all." Employees of Patagonia enjoy flex work hours, an on-site day care center, and parental leave for both mothers and fathers. Chouinard says that he wants employees to feel secure enough at home so that they can be creative at work. And, he says, Patagonia doesn't provide the flexibility and the benefits because they are "nice." It provides them because it's good for the company.

The turnover rate at Patagonia is 4.5% in an industry where the average is between 20% and 25%. When the average cost of hiring and training an employee at Patagonia is around $90,000, strategies devised for retaining seasoned employees just make good business sense.

- Live for Life at Johnson & Johnson represents a longstanding commitment the firm has to the safety and health of not only its employees but also their families. The J&J credo states that the company has several responsibilities: The first is to the customer—to deliver high-quality products at good value. Second, J&J is accountable and responsible to its employees and their families for providing them with a sense of security in their jobs, along with safe and orderly working conditions. Third, J& J has a responsibility to the community in which it operates to respect the environment and to be a good corporate citizen. J&J believes that if it does a good job with its first three responsibilities, the shareholders of Johnson & Johnson should come out O.K. The J&J credo seems to be working: Over the last 10 years, the shareholders have enjoyed over 20% compounded annual returns.

Source: Adapted from The White House Office of the Press Secretary Conference on Corporate Responsibility Panel 1, May 16, 1996, Georgetown University, Washington D.C.

The Body Shop: A Study in Values

Three decades ago, Anita Roddick founded The Body Shop with a borrowed 500 bucks. Today, she remains at the helm of an international success: 1,200 retail outlets in 45 countries and worldwide sales of $795.2 million. We had the opportunity to talk to Roddick, and what we discovered is that she mixes business with social responsibility with as much care as the skin creams and lotions her company sells.

Consider a corporation that:

1. Views volunteering in Romanian orphanages caring for AIDS-afflicted infants as the best teacher of what it means to serve. The Body Shop has sent more than 300 employees over the past six years to do just that. As Roddick proclaims, "It is a wonderful way to instill in people the very essence of service, whether it is to a customer over the counter or to a child." The experience, she says leaves a lifetime impression on the employee, who returns with a zest and a passion that only contributes further to the shared values of the organization.
2. Publishes an environmental and animal protection report and a social statement for customers and investors alike. An independent firm interviews employees, suppliers, and donor recipients to make certain that the company is meeting its ethical standards.
3. Incorporates social responsibility into every aspect of the company, from its products to publicly championed projects. Roddick firmly believes that business has a role to play and a debt to share on social issues.

Anita's latest project was the founding of an educational organization called The New Academy of Business. Described as a business school with a social conscience, it will offer courses in social and ethical auditing, strategic marketing, human rights and international trade, and spirituality in business. She says business should not be a "financial science, although we have turned it into that. Business is about people—that exquisite place where buyer and seller come together. Business is a place where relationships are born, and relationships are all about the heart."

Interview with Anita Roddick, Professional and Business Women's Conference, May 1996, Santa Clara, California.

WE CARE HOW YOU TREAT YOUR EMPLOYEES

A business is rightly judged by its products and services—but it must also face scrutiny and judgment as to its humanity.

D.J. DePree, founder, Herman Miller

At the risk of exaggerating to make our point, would you want to be in a personal or even a business relationship with a guy who regularly berates the people who work with him? Probably not. Most of us have winced when we've seen a tyrannical boss in action, spreading fear and loathing throughout the organization. The Abusive Boss is a staple of comic strips, cartoons, and sitcoms (Dagwood Bumstead and Mr. Dithers, along with George Jetson and Mr. Spacely come to mind) but, in real life, it's anything but funny.

The discomfort is not necessarily because of any personal involvement with the people subjected to the boss's tirade. More likely, the reason a person will not commit to such a relationship—unless it's an absolute career necessity—is because of simple humanity, caring about other people and how they're treated. Given the choice of befriending the guy who goes ballistic at the drop of a hat and a more fair and moderate boss, most everyone would pick the latter.

That's our inclination too. So, if and when we hear stories about how your company is difficult to work for and how tough your employees have it, we'll tend to find some other company with which to form a relationship. In these days of customer activism, a bad reputation in this critical area is almost as damning as a bad reputation for product quality or service.

On the other hand, we'll see stories in the paper and in *BusinessWeek* about Hewlett-Packard, Johnson & Johnson, and 3M, and other companies that regularly make the top 10 list of companies employees like to work for most, and we're almost always pleasantly predisposed toward them. Call it optimistic thinking, but we can't help but believe that the company that goes out of its way to treat its own people well will do the same for us.

Letter From Herb Kelleher About Southwest Employees

To Our Shareholders:

In fourth quarter 1994 and first quarter 1995, our year over year earnings were down substantially as the proud and beloved People of Southwest Airlines defended themselves against simultaneous assaults from Continental Lite and the United Shuttle. Because of my supreme confidence in the dedication, martial vigor, and extreme valor of our People, I predicted in our 1994 Annual Report that Southwest's fortunes would begin to recover in the second quarter of 1995. I also said that: "While a number of other airlines may attempt to imitate Southwest, none of them can duplicate the spirit, unity, "can do" attitudes, and marvelous *esprit de corps* of the Southwest Employees, who continually provide superb Customer Service to each other and to the traveling public. Just as the past has belonged to Southwest because of our People's goodwill, dedication, and energy, so shall Southwest seize the future!" ❤ As of today, Continental Lite has ceased to exist and the United Shuttle has substantially receded from Southwest's Oakland, California, markets. Because of our Employees' indefatigable efforts, a 72% year over year decline in first quarter 1995 profits has been transmogrified into a 1995 record annual profit of $182,626,000 ($1.23 per share), a 2% increase over the $179,331,000

($1.22 per share) of 1994. ❤ Our fourth quarter 1995 earnings of $43,359,000 ($.29 per share) made a very substantial contribution to our 1995 "turnaround," as they exceeded 1994's $20,343,000 ($.14 per share) by 113%. ❤ Despite the adverse cost impact of the recently effective 4.3 cents per gallon federal jet fuel tax, accompanied, as well, by thus far moderate increases in jet fuel prices, from this early vantage point we presently anticipate, barring any unforseen and deleterious external events, that our first quarter 1996 earnings will substantially exceed those of first quarter 1995. ❤ During 1996, we anticipate adding 20 new Boeing 737-300s to our fleet and removing three older 737-200s therefrom. Our net increase in available seat mile capacity is expected to be approximately 13%. As of this writing, our just inaugurated service to Tampa and Ft. Lauderdale is already producing average daily load factors in excess of our system averages, and we are, therefore, optimistic that Florida will be a successful addition to our growing route system. We will inaugurate service to Orlando in April 1996, and we currently plan to devote at least eleven of our twenty 1996 aircraft deliveries to our new Florida markets. ❤ As American Airlines has "dehubbed" Nashville, we have been steadily adding replacement service, and Southwest is now the largest Nashville air carrier in terms of daily flight departures. In 1996, we will inaugurate nonstop service from Nashville to Tampa and Orlando and one-stop service to Ft. Lauderdale. ❤ After beginning 1995 with a truly dismal first quarter earning performance, I am especially pleased and extremely happy to be able to report to our Shareholders 1995 annual financial results slightly improved over those for 1994. How was this "miracle" of 1995 accomplished? It was achieved through the fighting spirit of the marvelous People of Southwest Airlines. They never give in and they never give up; that is why they are my heroines and my heros!

Most sincerely,

Herbert D. Kelleher
Chairman, President, and Chief Executive Officer.
January 27, 1996

P.S. Our People are also a heck of a lot of fun to be with!

Along those same lines, we're also convinced that the way a company treats its employees will filter down to the way those employees treat us. In fact, we've seen proof of it time and again. Whether it's in the family, in the classroom, or on the job, people in a given social setting will model the values and behaviors they see.

Community and Values On-Line

In July of 1996, America On-Line launched a new area, consisting of a chat forum on corporate and social responsibility, information on how its customers can link up to nonprofit and community organizations throughout the world to volunteer their time and services, and a detailed listing with background information on the top 10 firms in the United States considered to be the most socially responsible.

According to Marc Aguierre, AOL's manager of community care, the new area was an overwhelming success. AOL was inundated by customers' kudos and requests for information. Plans are under way to expand the on-line area to include products and services from socially responsible companies in the AOL on-line store. In the future, a consumer will also be able to access a database loaded with information and sources about the socially responsible actions and investments of publicly held organizations.

When we enter into a relationship with you, we expect to be treated individually and treated well. And it is our belief that, if you don't treat your employees well—and as individuals—they'll pass that same discourtesy down to us. And, in that case, thanks—but, no thanks.

PART THREE

Building Relationships One Employee at a Time

The most profound effect of this (digital) technology revolution...will be that it will make the differences among people more important. As technology adapts to people, our ability to unleash human potential will be the primary source of competitive advantage.

—Kevin Kelley, founder, *Wired* magazine

In the past, except for occasional periods of financial downturn, there was always more demand for products and services than there were companies to supply them. No matter how many competitors cropped up, there always seemed to be more than enough customers to go around, with enough purchasing power to keep cash registers ringing. As a result, there was more emphasis on pushing products out the door than on innovation. Repetition was prized over creativity. Speed was held in higher regard than excellence. And risk taking seemed a needless exercise. Not much by way of exceptional effort or initiative was expected from most people and, consequently, not much was given. Individuals put in a reasonably good day's work for a reasonably good day's pay, augmented by the promise of lasting employment.

In the One Size Fits One future, supply, in most cases, will exceed demand. And, even if that isn't the case, the number of good customers available for a loyal relationship (i.e., those customers not already committed to another company) will be far smaller than the number necessary for many companies to succeed. Innovation, creativity, excellence, and the ability to respond quickly will be held in high regard by customers. And employees at every level will be called on to provide them. Work that is repetitious or mindless will most likely be automated, as we will no longer be able to waste as much of the human talent and spirit as we have so frequently done in years gone by.

SIMPLE PROCESSES/COMPLEX JOBS

In the past, in the mass-production model, simple, narrowly defined jobs filled by people who required neither technical nor educational sophistication led to the need for complex, sometimes arcane processes in which employees had no sense of the big picture and the left hand rarely knew what the right was doing; what's more, no one seemed to care. While designed to meet the needs of companies, not customers, and short on overall efficiency, these processes provided control over the workforce, made all but the most basic training unnecessary, and enabled the easy and inexpensive replacement of one worker with another. Employees were, in fact, frequently the least critical, least considered element in the process. In the mass-production model—and its antecedents—focus was on compliance, not commitment; on doing what you're asked, not what you think; on money and trinkets as motivators rather than the work itself; on the accomplishments of the individual laborer, not the work group;

and on internal competition rather than internal cooperation.

In the One Size Fits One world, the mass-production model won't work. Nor will the practices that underlie it. Complex relationships absolutely cannot be built by employees who, are forced by process design to be simple-minded. Instead, in place of these processes will be those that are streamlined and simplified—made possible, in part, by technology and, in even greater measure, by the fact that it will no longer be necessary for processes to control individuals or compensate for their lack of training. On the contrary, in the future, highly trained individuals working as independent and empowered decision makers, innovators, and continuous learners will handle the complex aspects of building ever stronger relationships with customers while processes support them in their efforts. In the One Size Fits One world, in other words, people will reign over processes rather than processes over people.

ACKNOWLEDGING THE KNOWLEDGE WORKER

In the past, conventional wisdom had it that most of the world's so-called knowledge workers were professionals: doctors, lawyers, accountants, executives, and so forth—people who worked more with their heads than their hands. Unlike most people who toiled in corporations or in government offices, these individuals didn't have jobs, they had responsibilities. Rather than being given specific tasks to perform, these knowledge workers were held accountable primarily for results. Did the patient get well? Was the defendant found innocent? Did the bottom line improve? Nor was their performance judged on the successful completion of a specific function: taking blood pressure correctly, making a good legal argument, adding debits and credits without making a mistake.

In the One Size Fits One future, where there is a direct correlation between how much you know about your customer and your company and the strength of the relationships you create, every worker will have to be a knowledge worker. Jobs, and the relative security they brought in the past, will be replaced by a more dynamic, more demanding—for both employees and employers—*employment contract,* where opportunity, freedom, and success will be afforded those individuals who take personal *responsibility* for bringing value to the organization and adding value to the work they perform. No longer will employees have job descriptions outlining the tasks they are to perform. Instead, they will be expected to do whatever is necessary to achieve the single overarching outcome that will determine the success or failure of their organization: creating loyal customer relationships. And, rather than judging their success or failure by their individual tasks—and irrespective of the larger result—employees will be able to gauge their value to the enterprise by their ability to create and nurture these relationships *one customer at a time.*

If this is what the future really holds—and there is every indication that it is—then how do we get there? How are we to make the transition from the demand-exceeds-supply, complex-process/simple-job, dumbed-down-worker past to a future where there are fewer good customers to go around and more and better companies competing for them, where processes are simplified and jobs are complicated by the need to form relationships, and where not only is everyone a knowledge worker but a knowledgeable one as well? How can we rid ourselves of our old, tradition-bound ways of controlling employees to unleash the potential of the workforce? Not, incidentally, that we have any choice. The die has been cast. One Size Fits One is on its way, whether we like it or not, leaving our leaders with little alternative but to deal with it.

THE CHALLENGES OF LEADERSHIP

In the One Size Fits One world—and more than ever before—our leaders must work to create a culture where the focus is first and foremost on customers and on the relationships we have to build with them—a culture where loyalty to the customer supersedes all other loyalties. To this culture, leaders must bring a vision—their own and the collective *shared* vision of everyone in the workforce. No longer can this vision be imposed from above as if by divine will, but it must be arrived at mutually by all those who would follow it...or they *won't* follow it. Leaders must also concentrate their efforts on building a day-to-day work environment where values are practiced on a consistent basis and held in the highest regard and where risk is not only tolerated but encouraged.

Finally, in order to unleash the full potential of the workforce, leaders must fight against the preconceived idea shared by many that employees don't want responsibility; they don't want to know more, learn more, be accountable for more. This will be a difficult battle, however, because there is more than an element of truth in this notion. How so? Because in the mass-production, simple-job/complex-process model, responsibility usually involves risk taking—and sometimes failure—which employees have learned must be avoided at any cost, even if it leads to mediocre performance and lackluster results. In the One Size Fits One world, however, mediocre performance and lackluster results are untenable, and risk and failure are part of the territory—not punishable by anything worse than the opportunity to try again, using what was learned the first time to improve our chances of success in subsequent attempts.

To those of us doing business today, the value of unleashing the human potential of our co-workers is hardly news. If our rhetoric is any indication, we all know how important employees are—particularly those on the front line—to building the strong, loyal relationships we must maintain with our profitable customers in order to keep from becoming the first dinosaurs of the 21st century.

We know this, whether we're delivering health care, writing software, serving fast food, or providing beauty care. We know that, if we're to thrive in tomorrow's One Size Fits One world—where relationships are king and where products and services will blend and the mousetrap will be only as good as the path that leads us there—we'll need more than a compliant workforce that checks their brains at the door and puts in their time. On the contrary, we'll need loyal employees committed to building that relationship, sharing a collective vision for doing so, and creating exceptional value for each individual customer along the way.

Also, as we're often quick to note, motivated, involved, and committed employees can result in a lower cost of doing business, higher productivity, substantial process improvement, faster service innovations, better customer information, new employee referrals, and much more. Lately, too, many of us are acknowledging, even advertising our commitment to employees: "Our people," we claim, "are our most important asset." In fact, just ask any of us how important it is for our organization to have the day-in and day-out commitment of the men and women on the job, and chances are excellent that our answer will be short, concise, and to the point: "Absolutely critical."

IT DEPENDS

But, if the commitment of our co-workers is as critical as we say it is, how do we win that commitment? What will keep Employee X or Employee Y loyal, creative, and excited? To these questions, we're often slower to respond. And when we do, our answers are almost always identical to those surrounding customer loyalty and commitment: "It depends," we say. It depends on the individual employees and who they are. It depends on their needs and expectations. It depends on what turns them on.

In other words, just as with customers, we can't generalize about what makes employees loyal. Every employee is different. Every individual relationship we form is unique. What works for one won't necessarily work for another. As we've learned from our relationships outside the organization, there's no simple, one-size-fits-all—or even one-size-fits-many—way to win loyalty and commitment. On the contrary, if winning the individual commitment of our co-workers is what we're genuinely after, we don't have any real alternative: Only One Size Fits One will do.

THE QUESTION IS...

We say it. We know it. We've *experienced* it...as employees ourselves. The fact is, as human beings, we're all looking for—we all actively seek out—One Size Fits One treatment. We appreciate being treated as individuals; we hate being treated as part of the mass. We like being listened to, understood, respected; we're angry when our requests are trivialized or ignored. We like feeling special; we can't stand the notion of being thought of as unimportant.

The question is, if we genuinely believe that employees have the same human needs as we do—in private life as well as on the job—and that the only way to win their commitment is by treating each of them as individuals, are the people practices we have in our organizations up to the task? Can we get to where we say we want to go with the systems we have in place? Is what we *think* about employees consistent with what we *say* about—or to—them? Pose this question to virtually anyone with a pulse who works in a medium- to large-sized organization and you're likely to encounter skepticism. Our words say that we're ready to put aside our mass mentality in dealing with employees and take the leap into a One Size Fits One world. In fact, most of what we *say* indicates that we've *already* taken that leap, and that we're well under way. In many instances, however, our actions suggest otherwise.

A LOOK IN THE MIRROR...

It's time we face the facts. While company after company claims that its employees are its most important asset, the majority of employee policies and practices simply don't bear this out. If, for example, we genuinely believe that employees are our number 1 asset and that the only way to win their individual commitment is by treating them as individuals, how can we expect them to take us seriously when we regularly create or perpetuate one-size-fits-all people practices like job descriptions that narrowly define the scope of work and rarely consider the scope of the individual's real talents and personal aspirations? Or performance appraisals that group individual employees into one of five boxes, pretending it's fair, pretending it's objective, and pretending the employees will learn and benefit from the experience? Or compensation systems that focus on getting people to do what we want rather than improving the process or encouraging learning and growth?

If we really cared about employees and were serious about them as individuals, wouldn't we develop a growth path for each of them that encompasses what that person wants or needs from his or her tenure with us? After all, if employees can't attain their personal goals at work, how can we expect them to commit to the job? If employees are truly our most important asset, why is it that the training budget is the first thing cut when money gets tight? And why is our first concern in developing a policy *not* usually whether it's good for the workforce but whether it's legally defensible? If we genuinely wanted employees to give us their best, wouldn't we do our best to ensure that they have the skills and knowledge necessary to do so? Failing that, wouldn't we at least want to know if they are married, have a family, have problems or concerns away from the workplace that we could perhaps help with, to win their respect and trust as well as their commitment? To treat them as human *beings* with individual emotions and concerns and not just human *resources?*

The problem today is that it's a rare company, and an exceptional leader, who dares to devote the time and make the effort to form the human relationships with coworkers that lead to the commitment and to the unleashing of human potential we all say we're looking for. In place of these relationships—particularly in larger companies—we tend to create one-size-fits-all systems to "handle" employees—practices and policies that are clearly throwbacks to yesterday's mass mentality.

If we are serious about meeting the challenges of a One Size Fits One world, we must look in the mirror and learn from our past efforts. We must make unleashing the potential of people a strategic imperative. We must *commit* to finding a better way.

Getting On With the Job

The management sciences—statistics, data analysis, productivity, financial controls, service delivery—are the things we can almost take for granted these days. They are subjects we know how to teach. But, we are still in the Dark Ages when it comes to teaching people how to behave like great managers, somehow instilling in them the capacities such as courage and integrity that can't be taught. Perhaps, as a consequence, we've developed a tendency to downplay the importance of the human element in managing. But the only people who become great managers are the ones who understand in their gut that managing is not merely a set of mechanical tasks but a set of human interactions.

—Thomas Teal, *The Human Side of Management*

WHERE DO WE BEGIN?

How do we go about getting "on with the job"? How can we learn to unleash the potential of the workforce to add value in a One Size Fits One world? How do you build an environment in which people will seek responsibility, search for better answers, and be willing and able to give their all? How *do* you motivate employees?

Before he died in 1963, M.I.T. professor Douglas McGregor repeatedly asked students and managers this very question: "How do you motivate employees?" In posing the question, he was not looking for a specific answer because he believed that a simple answer did not exist. Rather, he was attempting to get each person to articulate a *hypothesis*, a theoretical guess about what might entice people to make a maximum contribution at work.

In his attempt to help managers to articulate hypotheses, McGregor frequently came up empty. And the theories that people did put forth, theories like "Reward success/punish failure," often didn't jibe with the personal beliefs about human nature these people professed. If managers and students had theories, in other words, they were usually incomplete, frequently unexamined, and, in some cases, riddled with internal contradictions.

What McGregor encountered nearly four decades ago is probably not much different from what he might encounter today. Despite the fact that most managers agree that leading people is their primary job, most have not developed a hypothesis about how to go about unleashing the potential of people. And, if they have, it's more than likely that they haven't subjected this theory to rigorous examination. What *has* been examined—and with great rigor—are the numbers: If the numbers are good, the thinking goes, so is the theory; if they aren't, it's time to try another.

Whose Job Is It Anyway?

More than 40 years ago, it was suggested that human resources be divided into two separate departments. One department would be charged with the administration of pay, promotion, fringe benefits, and so forth. The other would have the responsibility for finding new and creative ways to unleash the power of people in pursuit of organizational objectives.

It was clear to these commentators that the single department was not working. They noted that the day-to-day pressures of dealing with current problems often took precedence over innovation. Also, many of the practices developed and administered by that department were inhibiting the development of new and creative ways of engaging people. They hypothesized that the new group would be freed from the pressure of defending the "old" and would have one mission—with no allegiances to the past.

The idea never took hold 40 years ago, but perhaps it is time to dust it off and think about the possibilities. It is the rare company in which the human resources department believes that its primary role is to unleash human potential. Most still talk about that role but fill their calendars and days defending the organization and administering yesterday's policies. We need both—but maybe not in the same department.

We are, by and large, a practical people. We prefer practice to theory and action to exploration and, at times, have little patience with ideas that cannot be practically applied *now,* during this quarter. In many cases, our predilection for action has served us well. But, in our efforts to create a work environment in which people can excel, our aversion to theory and to examining the consequences of our assumptions has often led us to create or perpetuate practices that, rather than motivating people, undermine them—rather than unleashing potential, restrain it. If we are to capture the creativity, imagination, and best efforts of people, then, we have no choice but, first, to "get theoretical," and, then, to explore the effects our theories have on the men and women we are counting on to carry the day in the One Size Fits One world. In other words, without theory, even our best efforts are bound to fall short.

HUMANS, NOT MACHINES

The major thrust of McGregor's writing about motivation was that managers seeking to unleash human potential must first abandon their oversimplified, mechanistic view of the workplace and learn to deal with the human side of the organization. No longer could employees be thought of as machine parts to be fixed, redesigned, or eliminated if problems arose. Instead, employees would have to be treated as individuals, in all their complexity.

Management by Gadgets

"I sat in a meeting a few weeks ago with a group of office department heads in a small company in Cambridge. They were considering the problem of getting people to arrive on time in the morning. What interested me was the way the conversation went in that group. One man said the solution was to install time clocks. Someone else said to take a little book and put it in a prominent place on a desk at the front of each department and require anybody who came in after starting hour to sign his name in the book and the hour of his arrival. Another man suggested that in his department he could arrange a turnstile at the door to the office in such a way that anybody coming in after 8:30 would ring a bell which could be heard within the whole department.

"These were serious suggestions made by a group of sophisticated supervisors who in general are doing a good job. They thought of the problem entirely in terms of the gadgets they could use to solve it. What they did not think of were the attitudes and the backgrounds they were bringing to bear on the problem. I saw in the discussion, although never expressed, feelings of this kind: 'Coming to work on time is something that people won't do voluntarily.'

"Those attitudes and convictions, always below the surface and never brought out into the open, were what

> was responsible for the suggestions they made for solving the problem."
>
> —Douglas McGregor, Speech to Management Forum
> E. I. duPont de Nemours Co., 1954

Putting McGregor's thinking into practice, however, as many of us know, is easier said than done. Even though most managers will quickly agree that people are more complex than machines, the systems that we inherited from the days of mass production and scientific management are often based on the assumption that people are no more than an extension of the overspecialized, overengineered process in which they work. Furthermore, the goal of these systems and the practices they spawned wasn't to get people to think, only to follow orders. And, with enough power, getting people to comply was not that complicated a task.

While these simple, mechanically based systems may be sufficient to obtain obedience to work rules and compliance to processes, they do little to improve the relationship with workers we will need to gain their commitment or stimulate a more learning-oriented culture. In fact, implementing and perpetuating these practices often distract us from the kind of experimentation and analysis required to understand how better to interact with individuals and groups of individuals. In the service of being able to control more effectively or manage more predictably, we our robbing ourselves of countless opportunities to learn how to be more effective leaders.

NO THEORY, NO LEARNING

Dealing with "living organisms" and their inherent complexity is the business of physical scientists, men and women who have been taught to take a very systematic approach to learning. In a highly disciplined manner, these scientists create hypotheses and then conduct a series of experiments to test these assumptions. Without a hypothesis to test, it would be impossible to learn systematically from the experiments that are conducted. If these scientists were only practical, action-oriented people who shunned theory and fled abstractions, we would know significantly less about our bodies and the universe. Fortunately, they are not. Instead, they accept complexity, chaos, and uncertainty because they know that that is the key to discovery, growth, and solutions of long-lasting utility and import. Besides, in the ever-evolving world of living organisms, complexity, chaos, and uncertainty are all there is. So, too, in the dynamic One Size Fits One world of business—though most of the systems on which we depend are predicated on exactly the opposite.

Organizational Renewal Begins with Personal Renewal

Most successful organizational renewal efforts begin with the personal renewal of that organization's leaders. Until the leaders of an organization are convinced that the methods of the past are no longer appropriate and that they must personally change, most renewal efforts are doomed to fail. It is through the leaders' efforts to learn—to rethink their thinking and to question their assumptions about what works and what doesn't—that the road to substantial improvement is opened.

Doesn't it make sense then to require that every leader routinely and systematically question yesterday's beliefs? Shouldn't we take the time to discuss our views and compare them with the views of our colleagues? As one person told us, "Thinking is just rearranging our prejudices." If we want to learn, we must be willing to question the views of others while having our own assumptions challenged.

If each of us is committed to discussing our answers to the following questions a couple of times a year, wouldn't we be more likely to articulate and scrutinize our assumptions about yesterday? And, having done that, wouldn't we be more willing to experiment with new methods tomorrow?

- What have I learned lately?
- In the last six months, what have I changed my mind about?
- When was the last time my assumptions were dead wrong?
- What is different about the way I think this year?
- What about my present mind-set have I found myself questioning recently?
- What have I learned this month that makes my actions last month seem less effective?
- Who thinks very differently than I do? What have I learned from them lately?
- How much time have I spent in the last month questioning the way I think and the structures I have designed to support improvement?

If we are to be successful as managers in this challenging arena, we must take a page from science and learn to evolve a hypothesis about how people are motivated and then test it rigorously against the practices we implement. In doing so, we must resist the temptation to embrace oversimplified models that fail to account for the complexity of individuals. We must take the time to understand people and, more particularly, the individuals that we work with, and we must build the skills necessary to learn systematically in complex environments. We must *learn to learn*—and, like the scientist, we must embrace the uncertainty that any profound and lasting learning necessitates.

Selection by Intricate Design

The people at Silicon Graphics Inc. (SGI) would be the first to tell you that selection is an art, not a science. They believe that even that best of selection processes fail, at times, to match people effectively with the "right company." However, at SGI, they clearly understand how important it is to their business to improve their batting average in the selection process—and they work at it.

First, they evaluate a candidate's education, experience, skills, and abilities. As a candidate, you don't get to square one without strong skills. Rarely, however, are these the deciding factors. In most cases, the following factors determine success or failure in the process:

Does the applicant:

- Have a demonstrated history of "seeking excellence"—not necessarily in a related field.
- Display self-confidence. A candidate who doesn't have confidence won't survive in the SGI system.
- Have strong creative abilities?
- Have the ability to absorb large amounts of information and use that information to solve problems?
- Possess strong emotional, physical, intellectual, and spiritual attributes and has demonstrated these qualities in past work experiences?
- Have the ability to articulate a vision for the future and a willingness to meld this vision with the visions of others?

What makes SGI's process so impressive is not its hiring criteria but the process through which these criteria were developed. SGI seems to understand its culture, its strengths, and its flaws, and is constantly evaluating its selection process to ensure that the people it hires can make a contribution to the business while they fulfill their personal needs.

LEARNING TO LEARN

In order to learn effectively in today's complex business environment, we must face the fact that we don't always have the answer and must be accepting, rather than defensive, when our ideas are challenged. This sounds simple but, in organizational settings, it rarely is. In too many organizations, admitting to not having the answer, especially about things as fundamental as leading people, is simply not career-enhancing. As one client of ours remarked, "Around here, if you build a ship and it sinks, you're better off just calling it a submarine."

This can't-afford-to-fail mentality has, in turn, spawned a host of antilearning behaviors that seriously hamper our ability to become better leaders. Rather than absorbing new information and finding new solutions to emerging problems, many of us have, in fact, developed a single script for motivating employees that we apply in as many situations as possible. Frequently, too, we are rewarded for our script as if, in implementing it, we understand its far-reaching consequences—even if we ourselves know we don't. Nevertheless, provided our script continues to be successful, we rarely stop to assess why. There is no theory, no experimentation, to our learning beyond whatever theory went into the creation of the initial script. Indeed, in most cases, learning ostensibly ceased after our first positive experience.

On the other hand, when we fail, we immediately stop to determine what went wrong and why. We test our hypothesis. Still, this is rarely the case, because our work environment doesn't tolerate too many ineffective tries. Truth be known, if we had failed more, we would have built better learning skills. We would have had to learn the consequences of our actions and to examine them in all their complexity. We would have learned more about leadership and the people we lead. We could not have pretended that we knew when we didn't.

Unfortunately, *except* in the critical area of unleashing the power of the workforce, most of us have been too successful. We've seen our scripts work again and again and have inured ourselves to the rigors of learning new and improved ones. We've been held accountable for financial results, mostly short-term, not the results of our ability to lead a motivated workforce. And, even when "people measures" have been used to evaluate our success—as in culture surveys—the numbers have been sufficiently vague or our failures sufficiently difficult to quantify as to make accountability difficult if not impossible. Sure, only 45% of the people trust management. But that's not bad compared to other organizations and, after all, some of our other divisions are worse. We simply have not had to face how little we know about people because our systems and our own adherence to them haven't held us accountable for it.

Generation X Teammates—Let the Revolution Begin

If we are the products of our environment, than the Xers among us are likely to be different teammates. Those of us who fit within this age category grew up in very confusing times, where turmoil in the home and the workplace was commonplace. Two-income families meant that a growing number of us found day care to be part of our daily routines more often, and for longer periods than ever before. Security in the workplace was oxymoronic from day one and, as one person told us recently, "Social Security has got nothing to do with me. It's a tax I pay so that the government can help support my mom when she grows old." For this generation to distrust large institutions is to be smart, and to live for today is common sense.

If we believe the sociologists, the resulting generation is likely to be at once more adaptive and more skeptical. Xers are likely to have well-developed interpersonal skills but will be wary of committing to any organization of substantial size. An Xer's need for variety and entertainment will reduce his or her willingness to work for prolonged periods of time in repetitive jobs. Independent, Xers will make more choices to leave sterile work environments for more inviting surroundings.

Maybe, the fact that this generation is so different will enhance its ability to help organizations renew. People in the X generation may be less optimistic about their futures but are more willing to work to be successful in the right environment. They yearn for a cause to believe in—but they will need to be assured that the organization can be trusted. They love to learn and, therefore, will be less tolerant of environments in which their talents are undervalued. They will not react to threats to their security as their parents did, because they are better equipped to deal with less certainty—uncertainty is all they have known.

This generation may help us deal with the inadequacies of our past practices because they will be less tolerant of them. Let the revolution begin!

If we are serious, then, about leading in a One Size Fits One world, we must become more disciplined in our attempt to learn what makes our co-workers tick. We must embrace the complexity of these issues and become avid learners rather than readers of old scripts. Of course, this is far easier said than done. To be a successful leader in this new world, we not only have to become students again and learn to learn, we have to continue learning throughout our career.

SYSTEMATIZED, NOT SCATTERSHOT

For most of us, the first step in the process of learning to learn is to develop a systematic approach to learning. Fortunately, the tools for this task already exist in most companies. Whether it is a five-, seven-, or nine-step continuous improvement process, a Total Quality methodology, or a "systematic" learning approach, most companies have been exposed to good methods for learning to deal with complex processes for some time now. Though saddled with a variety of monikers, each of these tools espouses basically the same methodology for learning:

1. Identify the gap between what is and what should be
2. Identify the probable cause of the gap
3. Research possible solutions
4. Test new actions
5. Measure the differences
6. Begin the process again.

Interestingly, all these learning tools are almost identical to the scientific method most of us learned in high school, with one critical exception: The scientific method demands a hypothesis, which the scientist tests against in the course of an experiment; the theory is then updated as a result of the new data that has been gathered. By adding theory to the tools that exist in their organizations, most managers already have the wherewithal to improve the learning at their disposal. That's the good news.

The bad news is that most managers' penchant for action leads them to these tools but *away* from developing the hypotheses necessary to make these tools effective learning devices. Instead, most managers carom from method to method, tool *to* tool, and model to model until they get the results they're looking for. And when they don't, instead of trying to understand why, they frequently blame the tool, blame the consultant, or, worse, blame the workforce for the failure of the organization to respond to a given implementation. Rarely, in any event, do they blame their own sometimes incomplete and scattershot learning methodology. Indeed, rather than taking the time to better articulate and understand their own theories about what creates success and what doesn't—particularly when it comes to leading a motivated workforce—many managers have recently become preoccupied with learning the practices of other excellent companies.

If we are to learn effectively from our experiences (or the experiences of others), we have to have a theory. Without theory, it is too easy to overdose on anecdotes. Without a good understanding of our theories, it is impossible to understand what is specific to a given situation and what is likely to transfer effectively to a different system. Without a complete understanding of our assumptions, we cannot understand the relevance of the techniques and tools we choose and are all too likely to introduce one program, one initiative after another until the entire workforce is confused or cynical.

BUILDING AN ENVIRONMENT IN WHICH MOTIVATED PEOPLE CAN CONTRIBUTE

Earlier, we posed McGregor's question, "How do you motivate employees?" We'll return to McGregor for the answer: *"You don't.* Man is a living organism, not a machine." This belief in the relative "unmotivatability" of humans because they are living organisms led McGregor to a theory of motivation with which we concur and which informs the later sections of this book.

Stated simply, McGregor's theory is that most people are born motivated to pursue what they perceive they need and that, if we want a motivated workforce, we must build and continually modify an environment in which people can fulfill these needs while pursuing the goals of the organization.

Organizational Dog Biscuits

"Why is KITA not motivation? If I kick my dog from the front or the back, he will move. And when I want him to move again, what must I do? I must kick him again. I can charge a man's battery and then re-charge it and recharge it again, but it is only when he has his own generator that we can talk about motivation."

—Frederick Herzberg, 1967

Herzberg liked to ask his audiences to identify the simplest, surest, and most direct way of getting someone to do something. Inevitably, after a short discussion of incentives, at least one manager would suggest that the person should be "kicked," preferably psychologically, not physically. Herzberg often referred to this method as KITA—giving a person a kick in the you-know-what.

To Herzberg, KITA was not motivation:
"I have a schnauzer. When it was a small puppy and I wanted it to move, I kicked it in the rear end and it moved. Now that I have finished its obedience training, I hold up a dog biscuit when I want the schnauzer to move. In this instance, who is motivated—me or the dog? The dog wants the biscuit, but it is I who want it to move. Again, it is I who is motivated and the dog is the one who moves."

In these cases, Herzberg believed that all that was done was that the KITA (the kick) was applied frontally. The dog was "pulled" instead of being pushed. But, in either case, the result was the same. The dog was not motivated to do anything other than to avoid the punishment inherent in the kick or get the reward promised for compliance. It always bothered Herzberg when managers consistently thought that positive KITA—the promise of a reward—was motivation and that negative KITA was inappropriate. In fact, the results achieved from the application of either KITA were exactly the same.

Most of us have learned how to use "organizational dog biscuits" successfully to get work done. Pay for performance, Person of the Quarter, Employee of the Month, sales contests, and management incentive plans, to name just a few methods, are really organizational dog biscuits. Many of us have become well versed in the art of creating movement in the workplace, but building a work environment to which people bring their own generators has been more elusive. To make meaningful progress, we must be willing to stop relying on "biscuits," and we must learn not to confuse movement with motivation.

The obvious key to the successful implementation of this theory is *alignment*—of personal needs and organizational goals—and any discussion of motivation stemming from it must take this concept of alignment into primary consideration. This is particularly true if what we hope to achieve is more than mere compliance. The argument follows that, once this alignment is achieved, businesses can benefit from the natural tendency of employees to act to fulfill their needs and, as they go about doing so, their actions will be consonant with the best interests of the organization. The challenge then, for those who concur with McGregor's theory or derivations of it, lies not in motivating people but in *building an environment in which motivated people are willing to make a maximum contribution.*

Most discussions of people's needs begin with a basic understanding of the work generally associated with Abraham Maslow. Its central thesis is that human needs are organized in a hierarchy, with needs for survival—food and shelter, for example—at its base. At progressively higher levels in Maslow's hierarchy are needs for security and social interaction, with the highest level being the need to learn, grow, and reach one's potential. According to Maslow, as lower-level needs become reasonably well satisfied, successively higher levels become more influential in motivating behavior. Also, when lower levels, such as security, remain unsatisfied, less energy remains for high-level activity, such as learning, creativity, or building self-esteem.

Most Likely to Commit

People who need the job the most are the ones who are most likely to be committed.

NOT!!! At least not according to a study published in the *Academy of Management Journal* (February 1995). In fact, the results of this study suggest just the opposite findings: that those people who are more secure financially or who believe that they can readily find a comparable job are the ones most likely to commit to a cause that they deem worthy.

Hungry people are certainly easier to control because they cannot afford not to comply. In their minds, it is often perceived to be more important to keep their jobs or get promoted than it is to do the right thing in pursuit of a mutually held objective. For too long, many have confused compliance with commitment.

Commitment is a personal choice. If people do not feel that they can make the choice because they cannot afford to live with the repercussions, there is no real choice and, therefore, no commitment. Perceived security is a prerequisite to building commitment.

This is not to say that, when lower levels are satisfied, they are no longer an issue. On the contrary: Maslow noted that people seem to have an insatiable ability to become dissatisfied with what he referred to as "environmental factors." Even when survival is not in question and people are reasonably well paid, they usually want to be paid better. Also, it's a rare individual who is completely secure in his or her work environment—though, today more than ever, that insecurity may be well founded. As for our social needs, they wax and wane on the strength of our personal relationships and our participation with others in the organization.

Reasonable satisfaction in Maslow's construct does not equate to complete satisfaction. Rather, reasonable satisfaction is achieved when, in the perception of the individuals involved, environmental factors like pay and job security are adequately addressed *and* equitably administered. People who feel that they are adequately and fairly paid do not spend most of their days thinking about their salary unless other environmental factors lead them to do so. People who feel that their health insurance is basically fair will not be preoccupied with the specific terms of the insurance contract. When discipline is handled consistently, most people are able to manage the risk of failure without allowing that risk to distract them unnecessarily.

Even when the actual amount of pay and benefits is substantial, however, if people feel that they are unfairly administered, the demotivational effect can be substantial. Only when people feel substantially satisfied in terms of both actual need satisfaction and relative fairness can they begin to focus on finding ways to concentrate their efforts on learning and growth. Alas, many of our present practices have been designed in ways that make relative fairness nearly impossible.

It's a Fairness Thing

We recently conducted an experiment in which we asked a person to take on a special project that would include substantial evening and weekend work, offering to double her salary if she would become a member of the team. She agreed without a second thought. We then recruited a second person for the team but had to sweeten the offer to get her to take on the project. As a result, the second person's salary was 10% higher than that of the first. When our first recruit found out what her new co-worker would be making, she became upset and threatened to quit. We asked her if we hadn't offered her enough money. She conceded that the amount we offered was more than enough, but that it was unfair for her to make less than her teammate for the same work. "Why does what we pay her have any effect on you?" we asked. "It's a fairness thing," she said.

REWARD AND PUNISHMENT

Much of what we know about building motivation revolves around the nature and effects of extrinsic and intrinsic motivational methods, with the former being by far the most recognized method utilized today (and in the past as well).

When a person is said to be *extrinsically motivated,* that person is seeking a reward or avoiding a punishment that is external to the activity itself. An extrinsically motivated individual, therefore, does something because of what he or she might get as a result. "If you do this, you will get that" is one popular way this type of reward structure has been characterized. It has also been less generously likened by some devout antiextrinsics (to which, with qualification, we add our vote) to rats traversing a maze in order to get the cheese at the other end. Of the rewards and punishments inherent in extrinsic motivation, money is the most common—but fringe benefits, promotions, praise, criticism, discipline, recognition, and social acceptance are also prime examples.

Many of us have significant experience in using extrinsic rewards and punishments to influence behavior. Employee of the Month, Man of the Year, management incentive plans, pay-for-performance schemes, and performance appraisal–linked promotion and pay are but a few examples. In fact, much of our early training had behavior modification—that is, using cheese (i.e., rewards) to control behavior—as its basic tenet. We learned that, if we reward positive behavior, its repetition is more likely; that ignoring a behavior tends to discourage it; and that punishing a behavior will probably suppress it—at least as long as the threat of punishment is seen as real. And we were consistently taught that our job as managers was to ensure that specific rewards were linked, by those we managed, to specific organizational goals.

Sales Without Cheese (Incentives)?

"You can't effectively compensate sales people without incentives. Straight salary? They won't produce—at least not at acceptable levels. If there is one thing that I was sure of, it is that professional sales people want and need incentives."

—Rene R. Champagne, Chairman, CEO
ITT Educational Services Inc.

And that probably would have been the way it would always be at ITT Educational Services if the government had not passed regulations outlawing incentives in the sales process. After an unsuccessful lobbying effort to overturn the prohibition, ITT developed a strategy to change the way it paid its sales professionals—straight salary.

But a funny thing happened on the way to compensation reinvention. Without the focus on short-term-sales, salespeople began to focus on recruiting only those students who were most likely to complete the course of study. As the top salesperson told us, "At first I fought the idea. After all, a real salesperson always looks forward to the commission. But I can't tell you how nice it was to be able to do the right thing for the new student and the school instead of doing whatever it took to make my numbers."

Much to the surprise of the CEO, who had grown up in sales and marketing, a number of positive results ensued:

- Sales professional turnover decreased by nearly 50%.
- The sales organization began to focus on the organizational mission instead of being preoccupied with their personal compensation.
- Some salespeople did leave because they didn't like being part of the team. In most of those cases, it was best for both of them.

- Revenues actually increased substantially because of the increased retention.

Beware of those things that you "just know" are true!

Ironically, it is these ingrained teachings about behavior modification—about the use of extrinsic rewards and punishments to get people to do what we want them to do—that may be the biggest stumbling block to building a workforce committed to the job, not the reward that comes after.

BEYOND THE CHEESE

Intrinsic rewards, on the other hand, are inherent not in the cheese but in the activity itself: That is, the reward is in the achievement of the goal. Intrinsic rewards cannot be directly controlled externally although the characteristics of the environment in which the individual functions can enhance (or limit) his or her opportunity to obtain these rewards. Achievements of knowledge, skill, or autonomy; self-confidence and respect; the exhilaration that comes from personal growth; the satisfaction that comes from helping others or being socially responsible are examples of intrinsic rewards.

For a variety of reasons, most of us have learned far less about, and paid less attention to, intrinsic rewards than extrinsic ones. A key reason for our failure to do so is that it is often difficult to establish a direct link between the leader's creation of an intrinsically motivating environment and the enhanced ability of followers to thrive, grow, and better serve customers in that environment. For example, while the causal connection between a manager offering a reward for achieving sales goals and a salesperson working overtime to win that reward seems obvious to both the offerer and the recipient, it is virtually impossible to demonstrate the causal relationship of a salutary work environment to an individual's or group's increased sense of self-worth gained from implementing an improvement that is fundamental to the success of the organization.

The Scanlon Plan—More than a Compensation Plan

Two decades ago, when we were first taught about the Scanlon Plan, we learned that it was primarily a gain-sharing compensation methodology—and a good one. At the time, we wanted to know whether gain-sharing paid dividends. If we dangled this "cheese" in front of the workforce, would people produce more? Wrong question.

To understand the Scanlon Plan (now referred to as the Frost–Scanlon Plan) is to know that the plan is about raising productivity and the level of organizational performance through increased participation and coordination. The sharing of the fruits of improvements was just one small part of a systemic approach.

The plan was based on four assumptions:

1. Change is inevitable and is our only hope.
2. Education is the best investment in achieving change.
3. Current behavior is a consequence of previous treatment.
4. Every person and organization is in the state of becoming.

Variable compensation need not be an attempt to manipulate behavior extrinsically. Plans like that of Joseph Scanlon and Carl Frost's demonstrate how effective it can be to take a systemic approach to improving performance. Ask converts to the plan like

Harley-Davidson, Herman Miller, Lincoln Electric, Motorola, and others about the benefits. The foundation of the effort in every case is a belief that people can (and will) make outrageous contributions to the business in the right environment. This plan is about improving the environment and sharing the results of the effort.

As managers in an intrinsically rewarding environment, we can also lavish praise on employees, but no one but the workers themselves can provide the reward that counts the most: a genuine sense of accomplishment. In any event, most of us like to know the result of our leadership interventions and, without direct feedback, we can never be sure of the effects of our actions. We like the certainty of knowing, and we often avoid that which cannot be directly proven. We like to be in control, and we simply can't control intrinsic satisfaction beyond creating and controlling the environment in which it takes place.

It's About Profit

"We're a company that is about profit. It drives us personally. It drives us as a company. It drives our incentive programs. And it is satisfying to our owners."

So spoke the CEO of a billion dollar service company as he gave the keynote address to begin the national meeting of company managers. The number 1 objective of the company for 1997 is to "motivate" people to provide better service. What are the chances?

A second reason that some managers have failed to experiment adequately with an intrinsically motivating environment may be negative assumptions they hold about people and their work ethic: If managers don't believe that their people are capable of finding work interesting and exciting, they won't spend the time to experiment with different environments. Of course, in today's relatively enlightened work environment, most managers would resist saying outright that they don't believe people are capable of learning and growing on the job. The fact is that many managers are quick to identify certain members of their organization as people who actively seek responsibility and are capable of accepting more of it. But, often, these same managers are just as quick to note that far more people in the organization can't be trusted to behave as these leaders do, while still other employees simply want to get the maximum reward with a minimum of effort. Therefore, many managers note, given the present makeup of the workforce, creating an intrinsically motivating environment, one less reliant on the manipulation of extrinsic rewards, is impractical. In fact, with so few rewards in the work itself for most employees, these managers conclude, the only practical solution is to continue to depend on extrinsic rewards. At least in this way employee behavior can be controlled. At least, compliance can be achieved, even if it means forgoing commitment.

Delancey Street Inc.— The Power of Belief in People

"It won't work with the employees in our business." Over and over again, we have heard that refrain from people who don't think that intrinsic motivation is practical in their work environment. Well, meet Mimi Silbert.

All Mimi wanted to do was to create a couple of businesses, managed and staffed by convicted felons—people who had, in the past, clearly demonstrated the willingness and the ability to commit serious and violent crimes. She wanted these people to meet and serve San Francisco's residents and guests and to compete with some of the finest restaurants in the world. And, if that weren't enough, she wanted the Bank of America to lend her 40 million bucks to build the project. No problem!

Bank of America lent her the money (with full knowledge that the convicted felons would also double as the construction crew responsible for building the restaurant and residential facility on the San Francisco waterfront). The project was built, it won several architectural awards, and Mimi and her colleagues paid the loan back a year early. All without prior experience in building anything—business or construction.

Delancey Street Inc. has become a San Francisco tradition for great food and great service—compliments of those recovering felons. In addition, the printing business and the furniture and office moving business are thriving. Mimi and clan just broke ground for a new blues/jazz/café/bookstore which they hope to open next fall. And yes, it's the same construction crew.

Sure, none of this would have been possible without the consistent and inspired leadership of Mimi Silbert. But, as she was quick to note, it could not have happened without the creative, committed, and extraordinary efforts of the people who work there.

So, how do you motivate people like those at Delancey Street? How do you make the decision to trust those who have not been trustworthy in the past? How do you build skills in business where none existed before?

According to Mimi, there is no recipe, but it unquestionably begins with a strong belief in people. She told us that it takes a lot of creativity and know-how to be a drug dealer. These people are smart, and she always believed that most of them (70%) will do great work when given the chance, coupled with a supportive environment where there are people who care about them and take the time to teach them and where they live with the unwavering notions that they will be held ruthlessly accountable for their actions and that the values of Delancey Street are to be lived day in and day out.

Now, do you still think your work environment is tough?

What makes the extrinsic conclusion so seductive for far too many managers is that they feel more comfortable thinking in terms of meting out cheese as a reward for desired behavior than reshaping their thinking or rewriting their script. It's easy, it's surefire, it's relatively safe. With enough power, we can even predict people's reactions with some certainty. We can organize specifically for desired tasks. In short, extrinsic incentives fit our tendency to view the behavior of the workforce in mechanical terms. Relying primarily on extrinsic rewards and punishments reduces our risk, and it enables us to get short-term results without having to adjust the environment in ways that would challenge our assumptions. It's simple, and we like simple. And, because, in years past, the competitive environment did not require a maximum contribution from people, we survived. That was yesterday, however. Tomorrow promises to be different.

But regardless of tomorrow's marketplace demands that we change, there must, in any event, be a better way for us to exercise our leadership skills. People are not rats in a maze; they are capable of great achievements. We have all had experiences where people were challenged to do the outrageous and have exceeded everyone's expectations as a result. Why don't we think that that is possible day in and day out in our organizations? Why do we settle for so much less?

It's Not What You Do. It's Who You Are!

Undeniably, the way we act—our style—affects how people react to us. Our actions speak more loudly than our words and communicate volumes about who we are and what we think is important. That being said, however, it's possible that we may have placed too much emphasis on the study of leadership style in recent years. To believe that people react only to our style or that we can manipulate behavior by picking the right style can be misleading.

People want to know who their leaders are, whether they care, and the extent to which they share similar values and goals. The way we behave is an indicator but, when the behavior is perceived to be unreal, the net effect of choosing a leadership style—no matter how rational—can be negative.

We have seen people succeed with all types of styles, even ones that followers agreed were not the best in a given set of circumstances. But, in those cases, the people being led agreed that the leader was passionate, committed, fair, respectful, genuine, and caring. Although the style was less than perfect, people respected who he or she was and often said that they would follow the leader anywhere. Leadership style is important—when it helps communicate who we really are. When we try on a style that is not us, we fool no one.

THE DARK SIDE OF INCENTIVES

The problem with most extrinsic incentives is not only that they undermine employee commitment through their focus on rewards rather than work but that they actually interfere with creating the highly flexible learning environment that we all need if we are to succeed in the One Size Fits One world. Consider what happens when we use extrinsic incentives to modify people's behavior—the promise of money if certain goals are achieved, for example. There is significant research to show that, unless they are independently wealthy, most people will not only take the fastest, most direct route to the money but will often resent what they have to do to get it. By holding out the promise of money—and the threat of not getting it—employees are forced to scramble for it any way they can, often at the price of doing the best job possible. At the same time, harking back to Maslow, the threat to survival and security posed by the possibility of not getting the money drops the individual down the hierarchy of needs and robs him or her of the energy necessary to focus on the learning and creativity required to satisfy today's demanding customer.

This same dynamic comes into play with traditional performance appraisal. If I become worried that my appraisal may affect my promotion and, therefore, my family's ability to maintain its quality of life, I am again forced down Maslow's hierarchy from the highest level, where I do my best work, to a lower level, where I spend much of my energy helping to ensure my survival and security.

The fact is, we don't have to look far in most organizations to see the dark side of extrinsic motivation. Auto repair shops have long had incentives that paid managers for increased sales. When they feel that they can be honest, many of those managers will tell you how important that extra money is to their families and how risky it can be not to "upsell" in their industry. Again, the extrinsic incentive has the effect of refocusing people at a lower level in the hierarchy thus making learning more difficult—not to mention the effect such incentives have on the willingness and ability of these people to do the right thing for the customer.

Internal competition for extrinsic rewards can have a similar effect, particularly when the competition clearly defines winners and losers. For example, we have seen high-level executives expend considerable time and energy to win a Caribbean cruise, often doing things that they know are not in the best interests of their customers or their company—a cruise that, incidentally they could easily afford on their salaries but that they have to compete for (and win) to save face or elevate their standing within their organization. In any event, no one wants to be a loser in the eyes of his or her peers. People are social animals and they want to succeed; a loss knocks them down a level on Maslow's hierarchy, providing them with less time and energy to devote to higher-level achievements.

Here, as in the other examples we have cited, the "losing" individual's creativity in the service of the company and the customer is compromised in the name of management's attempt to control behavior through extrinsic reward. Given the demands of the One Size Fits One world, we believe that forcing employees down Maslow's hierarchy and compromising employee creativity—particularly in the service of today's demanding customer—is simply too high a price to pay for control. And that, as we move confidently toward that world, a choice has to be made: whether to continue to keep a close rein on the workforce in an environment that promises compliance at best or whether to create a new, intrinsically motivated environment where the work itself provides its own rewards and where people can fulfill their own needs while committing themselves to the organization's objectives.

ARE WE WILLING?

Are we ready and willing to face the following realities? That the problem with the workforce is that they don't respond to motivation but the methodology we have used to try to motivate them. That people are behaving much as we have asked them to. That their reactions to our manipulation of rewards and punishments are predictable. And that these same people, given the right environment, the right tools, and the autonomy to be successful, are capable of making great strides in finding new and creative ways to build relationships with our customers.

If we are willing, then we need to begin our journey by identifying those behaviors and systems that people in the organization perceive to be inequitable and we need to destroy them—even before we have a tried-and-true replacement. We need to formulate a theory that describes the type of environment in which people can learn, grow, and make a contribution. We must learn from our experiments and must resist the temptation to resort to "cheese" the next time things don't go perfectly. We must also accept the fact that we won't do everything right the first time or even the second. But, by persisting, we will learn and get better with every try.

Fortunately, accompanying us on this challenging journey will be the individuals who serve our customers, employees motivated to make a valuable contribution, men and women eager to do the best job they can. And it is our job as leaders and guides on this odyssey to provide them with that opportunity.

Overcoming a History of Low Expectations

"The difference between a lady and a flower girl is not how she behaves but how she's treated. I shall always be a flower girl to Professor Higgins because he always treats me as a flower girl and always will; but I know I can be a lady to you because you always treat me as a lady and always will."

—Eliza Doolittle to Colonel Pickering in George Bernard Shaw's *Pygmalion*

If we have high expectations of what people will produce, it increases the chances that people will produce at higher levels. Today, few people would argue with this premise—often referred to as the Pygmalion effect. Most of us, however, whether in our schools, organizations, or homes, have not translated our knowledge about higher expectations into practice. We simply don't treat people as if we expected a great deal from them—and, as a result, we often get what we expect.

Harvard University psychologist and researcher Robert Rosenthal conducted one of the most revealing studies regarding the power of expectations on performance in the South San Francisco elementary schools. In the study, reported in *Pygmalion in the Classroom,* Rosenthal randomly selected 20% of the student body and told their teachers that these children had unusual potential for intellectual growth. Predictably, the teachers had higher expectations for these students and treated them differently than they did other students in the classroom. Also predictably, the children responded. They learned at an accelerated pace and, by the end of the school year, showed significantly more progress than their peers.

Rosenthal had demonstrated what most of us believe to be true—that our expectations can dramatically affect performance. But what we have not asked enough—either in the school environment or at work—is why we often have such low expectations of people. After all, if most teachers expected great things from all students, Rosenthal's research would not have yielded dramatic results.

There is little doubt that high expectations have not been the norm in most business environments. It's not that we have not expected people to work hard. Most of them have. Most managers, however, have underestimated people's ability to think, learn, and make decisions. In a variety of subtle and not so subtle ways, people have been told that they are not trusted to make decisions or to control the processes in which they work. Narrow job descriptions coupled with political turf issues have curtailed people's ability to take on more responsibility. Incentive plans have communicated a belief that, without "cheese" (a reward), management believes that people will not give their best effort. Implicit or explicit forced-distribution performance appraisal has sent the message to many that they are not trusted to give honest feedback, and so the organization has mandated that managers grade a certain number of employees as average or unsatisfactory.

What do we see when we look people in the eye? If we do not see someone who is capable of great commitment, creativity, continuous learning, and effective decision making, then we probably will not treat those people in ways that communicate the highest of expectations. If we do believe in the potential of the people in our organization, then we must quickly analyze our policies and practices in order to rid our organization of structures that implicitly communicate anything other than the highest expectations. To paraphrase Henry Ford, whether we believe they can or they can't, we are probably right.

The Challenges of Developing an Intrinsically Motivating Environment for the One Size Fits One World

It has become trite to say that...industry—the economic organ of society—has the fundamental knowledge to utilize physical sciences and technology for the material benefit of mankind, and that we must now learn how to utilize the social sciences to make our human organizations truly effective.

Many people agree in principle with such statements; but so far they represent a pious hope and little else. Consider with me, if you will, something of what may be involved when we attempt to transform the hope into reality.

Douglas McGregor
"Adventure in Thought and Action" speech
M.I.T., Cambridge Massachusetts
April 9, 1957

There is no one "right" way to create an environment where people will be motivated to do the things necessary for an organization to thrive in a One Size Fits One world. There is no single recipe, no grand plan. On the contrary, every environment will be different. And the means to achieve it will be as varied and unique as the individuals who populate it. Nor will a truly productive environment ever be complete; rather, it will remain a work in progress.

As such, our success in creating and growing this environment will depend, first and foremost, on committed, open-minded people willing to discuss, debate, and experiment with ideas that might prove useful and, second, on a strategy rooted in the unwavering belief that there is always a better way.

The 10 challenges for building an intrinsically motivating environment discussed on the pages that follow are by no means the only challenges to be considered in transforming our hope for a more effective human organization into a reality. It is our intention, however, that they will serve as a springboard to further dialogue, a critical review of past practices, a reexamination of beliefs, and the impetus for positive change and significant growth.

Ten Challenges for Building an Intrinsically Motivating Environment

1. Creating a cause worthy of commitment.
2. Building core values that guide performance.
3. Committing to the truth and sharing a sense of reality.
4. Ensuring that every person has a meaningful role.
5. Increasing accountability, but not more of the same.
6. Developing ability commensurate with responsibility.
7. Building cooperation instead of internal competition.
8. Abolishing the corporate caste system.
9. Developing an optimistic, caring, and supportive environment
10. Building trust one employee at a time.

Many of the ideas and concepts presented here were initially developed and published in *Leadership and the Customer Revolution* by Gary Heil, Tom Parker, and Rick Tate (Van Nostrand Reinhold, 1995).

1. Creating a Cause Worthy of Commitment

Most leaders tend to view teamwork as a social engineering problem: take X group, add Y motivational technique, and get Z result. In reality, the most effective way to forge a winning team is to call on the players' need to connect with something larger than themselves.

Phil Jackson, Head Coach, Chicago Bulls, and author of *Sacred Hoops: Spiritual Lessons of a Hardwood Warrior*
Interview, CNBC, January 20, 1996

Over the last few decades, there has been a gradual undermining of many organizations and institutions that people have traditionally turned to for a sense of belonging. There's been a notable de-emphasis on the extended family, and even the nuclear family has been diminished by separation and divorce. Many fell away from religion when they saw its leaders not practicing what they preached. Many more became cynical about government when the self-serving actions of elected officials were revealed and when government no longer appeared to work. Longtime political party affiliations were also weakened as parties adopted will-o'-the-wisp platforms that were more a response to trends than a reflection of long-held beliefs. More and more of our traditional institutions have become harder to believe in.

Disappointment in our traditional institutions notwithstanding, people all over America are still looking for that worthwhile organization and cause to commit themselves to—and maybe looking harder than ever. If this weren't the case, why does one-third of the workforce volunteer more than three hours a week outside the workplace? Why do many of them work harder for that volunteer organization than they do for the one that pays them? Why would many of us die for a cause when precious few would die for money?

A Political Antidote— Goodwill as a Core Value

We recently visited a company whose corporate values include "the attribution of goodwill." That means that each person attributes the best of intentions to his or her co-workers. An employee who does not understand why a co-worker has done something must ask instead of assuming that the co-worker has a self-serving motive.

We were surprised at how living this value reduced internal politics. Instead of talking negatively about people in the organization, employees seemed to go out of their way to be understanding. Admitting mistakes seemed easier when people were willing to assume that whatever was done was meant to help the organization reach its goals.

People in this company understand that not everyone can be trusted. They know that there will always be those who abuse that trust. But they are unwilling to allow the actions of a few to dictate how business is conducted among the 98% of employees who try to do the right thing.

It's not for everybody. In this company, if you lose the trust of your teammates, they ask you to leave. Wouldn't it be nice to work in an environment where people do not talk negatively about their teammates and where trust is the norm instead of the exception?

These questions go to the very heart of the issue we confront in remaking the workplace into one that is intrinsically motivating. Collectively, we are a people seeking something powerful and meaningful to touch our hearts and minds. We are at our best when we are swept up by commitment and are working in the service of a larger goal. We're looking for a cause that fires our imagination and excites our spirit. For the organization that can satisfy these needs—particularly today when there is a dearth of organizations that can—an enormous opportunity exists.

People want a cause that distinguishes them from others. No one gets excited about being average. If we don't stretch our goals, offer a valuable cause, and give people an opportunity to realize their potential, chances are they won't give us the best they have to give. This doesn't mean that every biotech company has to commit to wiping out hunger and starvation on the planet, or that every retailer has to clothe the homeless. Less ambitious causes can make employees proud to be working for the organization. The biotech company can focus on leading the industry in developing environmentally friendly products that replace traditional pesticides, while the retailer can content itself with striving to offer the best service in town. The measure of a worthwhile cause is not how lofty it sounds to an outsider or how it assuages the social conscience or stimulates the competitive juices of the management council.

A good cause is one that excites employees, deepens their commitment, and lends meaning to their work.

Can I Be Proud of This Group?

We credit Max DePree, former chief executive officer of Herman Miller and author of several books, for many of the following questions. Over the years, we've added our own questions to the list:

- Can I be proud of this group and its mission?
- Does what we do make a difference to anyone?
- Can I achieve my own goals by following you?
- Can I reach my potential by working with you?
- Will I be challenged? Can I learn here? Are you an avid learner?
- Will you be honest with me?
- Do you care about all the members of your team?
- Can I trust you? Will you trust me?
- Will you value my talents and opinions?
- Do you value the differences in people?
- Do you have a sense of humor?
- Will I be proud to tell my family and friends that I work here?
- How have you prepared yourself to lead?

2.

Building Core Values That Guide Performance

It is really fantastic to me that one book after another will make a pious statement about this new development and about organizational theory and management theory and then proceed to say nothing whatsoever about values and purpose except in some vague way that any high school senior could match.

—Abraham Maslow, 1963

A company's values are its code of ethics, its behavioral framework. Taken together, they form a statement of what the organization collectively deems important or valuable—what it stands for. When understood and adopted by employees, values provide a context for action. Values can provide a sense of order without rules, reduce ambiguity without a detailed plan, and bring focus and coherence while allowing individual expression and self-determination.

Most organizations have a published set of values. They are usually well written and widely communicated...and are a significant source of frustration. Why? Because it is a rare company today in which people practice the values their organization preaches. Just listen to the grumbling at coffee breaks, read between the lines at the next employee meeting, or read the verbatims on the employee survey—the gap between our professed values and our values in practice is significant:

- If respecting the individual is a core value, why do we promote so many people who are technical experts but have lousy people skills? And why do we tolerate managers who get results but are frequently disrespectful to people in their work groups?
- If we value cooperation and teamwork, why have we designed so many structures that reward people for competing more than cooperating?
- If every employee is encouraged to seek excellence, then why do we still have so many uninteresting, narrowly defined jobs? Why is training the first thing to be cut when money is tight?
- If providing the best possible service to customers is a core value, then why do we make so many rules that make it hard for employees to meet customers' needs?
- If we have adopted a Total Quality approach, why are so few senior managers actively involved in the effort?

...and the list goes on. Our failure to measure the extent to which we live our espoused values has made it easy for many to overlook how often we do not "walk the talk." The result has been lower levels of trust, greater confusion and frustration, and less sense of community among employees in many organizations. People genuinely want their company to stand for something. And, when it does, it increases people's feelings that their organization is special.

The first step in closing the gap as it relates to values is to choose to get serious about doing so. Look at the values you've published. Just how committed are you about acting within this framework? If a person purposely acts in a manner inconsistent with these values, what is the likely consequence? Dismissal? Coach and counsel? For most, it's an easy decision when an employee steals from the company. But what if one of your core values is respect and a longtime supervisor is clearly and consistently disrespectful of an employee in front of the work group? When this supervisor keeps his or her job after acting disrespectfully, what is the message sent to others in the organization about the importance of core values?

We know it's difficult to be truly tough on values. But we wonder if we haven't let the pendulum swing too far in the "give 'em another chance" direction. If there isn't significant accountability for the clear and deliberate abuse of values, then there are no values. On the other hand, when you do take action, when people who don't live the values are asked to leave, there is usually little, if any, mourning.

If we're not going to live our values, we would be better off not listing them. The only thing worse than not having a set of guidelines to provide order to behavior within the system is to have a visible set of values that are situationally enforced or largely ignored. In these cases, employees are reminded daily of how little leaders in the company can be trusted to keep their word. Values must be every organization's North Star—an ever visible guide that employees can look to, in order to tell whether they are acting in ways that promote the organizational good.

To be effective, values must be:

- Profound enough to touch the hearts and minds of all employees, yet simple enough to be readily understood.

- Concrete enough to provide a useful framework for decision making.
- Pragmatic enough and sufficiently consistent with organizational structures to be reinforced in normal day-to-day activities.
- Communicated over time in every aspect of the business.
- Reinforced through accountability.

A Suggested Vision and Values Building Process

1. Augment your management team with representatives from other levels in the organization. Through dialogue, build a picture of the type of organization that you would like to create to meet tomorrow's challenges.
2. Ensure that the draft that evolves from this session describes how the company will relate to shareholders, customers, and employees.
3. After the first draft is completed, members of the management team should share it with their direct reports or a larger group. People should come to this meeting prepared to compare their personal visions and values with the initial draft. Are there more than semantic differences? Are there statements made in the draft that don't belong? Can the group commit to the draft? What changes would be required before the group could get excited by the challenges contained in the vision and values draft?
4. Comments and criticisms of the draft should be addressed, with changes made at the management level, and then communicated back to the group.

5. This process should be repeated throughout the organization, with challenges to the draft allowed to sift up the chain of command (if there is one) and with reactions and changes as a result of these challenges discussed as they arise.
6. Build a feedback loop to test the effectiveness of the overall process, to ensure that the feedback emerging from these dialogues is unbiased, and to keep skeptics at bay. Keep in mind that any process you adopt for building a shared vision will be only an approximation of the more effective one that can result from feedback. There are no substitutes for systematic learning and persistence.
7. Expect and encourage minor differences in personal visions. When personal visions differ significantly, check information-sharing practices to make certain that people have the information they need to make informed decisions; then, check training practices to ensure that people have the ability to understand the significance of this information.
8. Once a broad base of people commit to a shared vision and values statement, ask them to define their individual roles in building the organization and to devise a feedback mechanism they can use to measure their progress. Each person must then be given the responsibility to evaluate his or her progress over time, communicate key learnings, and build the skills necessary to meet future challenges.
9. Assign a team to evaluate progress and to make recommendations about present methods. This assignment of responsibility may make it more difficult for busy leaders to push their efforts onto the back burner. It is crucial to build in accountability for the management team.

3. Committing to the Truth and Sharing a Sense of Reality

I believe it is wholesome and essential to recognize candidly that reality is all we have to share. To share that reality appropriately is a genuine favor. Facing up to reality should be viewed as a genuine opportunity and accepted as a challenging responsibility. True leaders are seldom remembered for asking their followers to remain the way they are. We revere Abraham Lincoln not for perpetuating slavery but for ending it.

Carl F. Frost, *Changing Forever: The Well-Kept Secret of America's Leading Companies*

It's difficult to see reality, period! No matter how empathetic we are (or think we are) as leaders, no matter how open, the organizational deck is stacked against us. When we are liked, people often color the information they give us because they don't want to disappoint us and they don't want to complain. If they fear our reaction (either because of our past actions or our position), very little information about the less than positive aspects of the current situation is volunteered. Most of us have felt the pressure not to be "too negative" or to ensure that bad news is balanced with enough positive information—even if this requires coloring or recasting reality. The expression "You tell the boss" didn't evolve because people were reluctant to be rewarded but because they didn't want to experience the boss's reaction.

In most organizations, people are biased toward seeing things as they wished things were, not as things actually are. The choice to avoid the truth is not always a conscious one and is often quite subtle. Just look at the number of ways people in your organization are implicitly (and sometimes even explicitly) rewarded for making things sound better than they are or telling their bosses what they think their bosses want to hear. Bad news doesn't travel "up" quickly in most organizations, but it should. By selectively modifying reality, we falsely diminish people's understanding of "the gap" and, therefore, rob the organization of the sense of urgency and creative tension that will be required to focus and sustain improvement efforts.

Developing a shared sense of reality must be a strategic issue. It is exceedingly difficult to change reality if you don't see it or if there is no agreement on what "real" is. We must recognize the effect present structures have on our willingness and ability to see reality, and we must search out and eliminate those practices that make it safer or more rewarding to do anything but "tell it like it is."

4.
Ensuring that Every Person Has a Meaningful Role

You can't buy people's time; you can buy their physical presence at a given place; you can even buy a measured number of their muscular motions per hour. But you cannot buy enthusiasm...you cannot buy loyalty...you cannot buy the devotion of their hearts. You must earn these.

—Rev. Martin Luther King Jr.

It is critical that we provide the opportunity for learning and that we design processes that people can control to create value instead of processes that control people. A mind-set that results in employees being systematically denied training and information flies in the face of our expressed need for the well-trained, well-informed employee. Indeed, in today's competitive, information-sensitive world, we'll have to retrain experienced workers and continuously retrain them for more complex tasks.

Undertrained employees will not be able to analyze information reliably and make the real-time decisions that will be required. Uninformed, partially committed employees cannot be expected to innovate delivery effectively or ensure customer loyalty. On the other hand, if our organizations provide employees with a worthy cause, interesting work, sufficient training, and the opportunity to reach their potential, the chances are excellent that they will make the commitment we seek.

Call the Headhunter—I Need a Partner

"Call the headhunter. Tell him we need to hire another tire changer." The first time we heard this statement, we thought that we must have misunderstood. A headhunter to find someone to change tires in a tire store?

That's exactly what Barry Steinberg used to do as he helped his team build Direct Tire Services. He also used the headhunter to help him find the best mechanics and the best brake people. When he found these people, he provided them with the best education available, involved them in the management of the business (not just their technical specialty), and was consistent in trying to ensure that every person who worked in the operation became so competent that they could get a job anywhere.

And the cost? It must cost more to keep these well-educated people on the payroll, to pay the headhunter. The education must be expensive. He also pays these people 15% to 25% more than other companies.

It's an investment, according to Barry—and a good one. Only by having the best team can he afford to guarantee the subjective satisfaction of his customers, charge 10% to 12% more than his competition, provide large windows where customers can watch the efforts of the team, and ensure that the experience is honest, fair, and done right the first time.

To him, people are not a cost to be minimized. They are partners in the enterprise, and it pays. His returns to the bottom line (in case you were wondering) are double the industry average. And, by the way, he rarely uses headhunters today. He has no turnover.

Unfortunately, most jobs have not been designed to produce value for the customer while also providing an opportunity for the development of the employee. In fact, in a large percentage of cases, just the opposite is true. We've designed jobs and written job descriptions in ways that make sense from an efficiency standpoint but that do little to feed the needs of people to learn, grow, and be challenged. In the short run (quarter to quarter), the current system may even seem effective. However, as the organization increasingly calls on the intelligence of its people to build new capabilities, the price we're paying for these "efficiencies" is already too steep.

The Other 'E' Word

A friend of ours received a notice of auto insurance rate increase a few months back. She was shocked to find that her premium had been raised by more then $900. Upset and furious, she immediately called the company's headquarters for an explanation.

"You were in an accident," the clerk she was connected to told her. He was not her agent, nor was he anyone she had ever spoken with before. "Yes, but that wasn't my fault. The other driver was cited." There was a pause on the other end of the line. "I have your file in front of me," the clerk said. Then, after a few more seconds, he concluded, "You're right. We must have made a mistake. We're sorry for the inconvenience. Please disregard the notice."

When we heard this story, we were impressed with how easily the mix-up was resolved and that a front-line worker had the authority to make a $900 decision on the spot. So we called the company ourselves and spoke to one of its executives.

"Tell us about how you 'empower' workers," we asked.

"We don't use the 'E' word," he said. "We use the other 'E' word."

"Which one?"

"Education," he said. "We ensure that every person who has contact with a customer has all of the information they need to make a business decision." He went on: "We think it's a crime to try and decentralize decision making without giving people the tools to make good decisions."

The question is: Is it possible to create an interesting, challenging job for every employee? Can we really organize so that every employee has a chance to grow and prosper? And, more to the point, is it even possible to accomplish this without turning the organization completely on end? The answer is yes—provided that we make the change from the old-fashioned, factory-based, production-oriented mind-set to one that better accommodates flexibility and responsiveness and, above all, expects significant contributions from everyone in the company in both the planning and execution of their jobs.

Certainly, not all jobs are equally interesting, nor can every job be made to seem appealing. This is particularly true in inherently unchallenging, highly repetitive, low-paying, little-experience-required jobs, of which there are millions. However, even in most of these jobs, interest and commitment are possible for many employees if, in addition to the performance of the task, they are charged with the management of the process.

What would this entail? Here are eight responsibilities that are the foundation of process improvement efforts:

1. *Gather customer information.* In the normal course of business, employees should regularly ask

customers what they like or don't like about the company's products and services and what they'd like to see done differently. This data should be quickly analyzed by those collecting it and then be communicated throughout the organization.

2. *Help design the service delivery process.* Being involved in this design, of course, must be predicated on an understanding of the entire delivery process, not just the job that the employee performs. Everyone must have a substantial understanding of the implications of their work and have the ability to influence the design of the way that work is accomplished.
3. *Customize the process when necessary.* Why is it that some employees don't act in the customer's best interest when common sense often dictates a different, obvious course of action? Often, it's because many employees today still aren't trusted to make even the simplest deviations in the process and must get supervisory approval to do so. For an organization to be considered responsive in the future, this must change.
4. *Measure the quality of their own performance.* The primary benefit to be gained from measuring their own performance is that the direct, unfiltered feedback employees receive when doing so enhances learning. Whether employees are or aren't allowed to measure their own performance is often an issue of trust. Many managers believe that employees are unable to, and haven't been trained to, measure their performance effectively and won't identify areas in which they have weaknesses because of the risk. Wouldn't they be honest in their self-appraisal if fear were eliminated from the process?

5. *Identify disgruntled customers.* Often, the customers we lose are the ones who are marginally dissatisfied but are not particularly vocal about it. Most customers don't complain; they just move on. Customer-contact employees can often head off such losses if they are given the responsibility to seek out those quiet, unhappy customers and do something to alleviate the causes of dissatisfaction quickly.
6. *Find the root cause of service problems.* In most service companies, management controls the process, and only a few groups have been trained to analyze information effectively to determine the root cause of problems. Rectifying this shortcoming will require education and time. Many companies still claim that they can't "afford" to give employees time to meet for an hour or two at the end of every week to discuss what they learned that week and to set priorities for improvements in the coming week. To be successful in the future, they can't afford not to.
7. *Improve the process.* Innovation and continuous improvement of processes have not traditionally been responsibilities of most employees. There is a significant difference between performing a task and playing a part in managing a process. The former can get repetitive quickly; the latter is continually engaging.
8. *Have the power to routinely eliminate non-value-adding tasks.* In most organizations, many people are involved in some unproductive activities. Typically, only managers can choose to eliminate these value-subtracting tasks. Why? If we respect people's minds, why don't we trust them to get rid of activities that waste their time—or at least build a process that enables them to point out wasteful activities without fear of retaliation?

5. Increasing Accountability, but Not More of the Same

With the freedom to choose come the consequences.
—Max DePree

Of the many ways organizations are trying to improve the performance of their employees, increased accountability is one of the most important. In most cases, if people are accountable for what they do, they'll do it better. Accountability can help provide focus, communicate priorities, indicate serious commitment to an issue, create a sense of urgency and tension, and demonstrate to all those in the organization that its leaders are even-handed and fair. In addition, and perhaps most significantly, recent studies indicate that accountability may be the single most important factor in effective decision making. It's hardly a surprise that people who have to live with the consequences of their decisions tend to make better ones.

The wrong kind of accountability, however, can focus energy on the wrong kind of activities and can lead to the formation of habits that must be broken. For example, it is common for many managers to be largely accountable for pleasing their bosses or making their bosses look good to their bosses. Often, too, managers are held accountable for short-term improvements in performance measurements that are achieved by mortgaging future opportunities or for meeting agreed-on performance targets that may represent an improvement over yesterday's performance but neglect the possibility of far greater improvement.

Nonperforming Whales, Act 1

A colleague of ours visited the pool where killer whales were being trained at a large amusement park. When the trainer asked if anyone knew how killer whales were trained, our colleague raised his hand.

"You feed them a few of the fish when they do well and, when they don't do well, they don't get the fish."

When he saw the trainer smile, he knew that he was in trouble and wished that he could take his words back. No such luck.

"You mean you want me to punish them—not feed the nonperformers, make sure the whales are good and hungry, and then get into the tank and swim with them? I don't think I want to be part of that experiment. Thank you very much," said the trainer.

Whales have to be fed. You can't train whales to learn if they are preoccupied with their hunger (lower level needs). That was an important lesson.

With people, as with whales, we can learn only when we are not preoccupied with our survival and security. The good news with whales is that such a strategy is life-threatening. Unfortunately, with people, we often manipulate them, often using the organizational equivalent of the fish without fear of immediate harm and, therefore, do not feel the need to find a better way.

Meanwhile, at lower levels of the organization, employees are frequently held accountable for complying with standard procedures and doing their job as it was designed to be performed instead of seeking out new and more efficient ways of doing the work, innovating the process, and better serving the customer. Left unchanged, these accountability practices wed us to the past. They make it more rewarding (or at least less punishing) to do things as they have always been done. If, however, we want to provide people with a challenging role in a changing organization, we must design accountability so that it supports that role and encourages that change. The focus needs to be less on immediate (and sometimes meaningless) results and more on continuous (even unreasonable) improvement and learning. As legendary UCLA basketball coach John Wooden advised, "Don't measure yourself by what you have accomplished, but by what you should have accomplished with your ability."

The Recovering Waitress

When we first met Diane, that is how she referred to herself. She described how, after three decades of waiting tables, one of her customers, the CEO of Krueger International, offered her a job. By her own admission, she knew nothing about making chairs, but accepted anyway. "And now," she was quick to point out, "we have grown from shipping $2 million worth of chairs a week to shipping $14 million worth—with the same number of employees."

When we asked her to tell us more, we were amazed. She described how Krueger International (KI) was growing because the company had developed the ability to manufacture custom chairs profitably—even if the customer wanted only one. Diane went on to describe how it took between 30 seconds and two minutes to change colors on a paint line and how certain sections of the production line had less than three complaints about their work in the last 12 months. Scrap rates? She knew them by heart. And that was just the tip of the iceberg.

The challenges encountered when KI moved to self-managed work groups? The benefits of an effective suggestion program?

In her words: "It's not about money for people. It's about pride. Some companies have come and tried to copy our system, but they have missed the point. Yes, we award people tokens for suggestions. Yes, people use those tokens to try and win bigger prizes. But it's not about the prizes for us. It's about making KI successful. We tell people that but, for some reason, they don't believe us."

If Diane had not told us that she had no special education and that she had spent most of her life as a waitress, we would have sworn that this "MBA" was part of a management team reinventing the contract furnishings business.

"It's amazing what people can do when given a chance," she added. "KI gave me a chance to make a difference. I'll be forever grateful. It changed my life."

Most organizations work from the assumption that management knows best—that managers ought to have control until a convincing argument can be made to the contrary. If lower-level employees (or teams) seek greater authority, they must ask for it, and then demonstrate that they deserve it and will use it wisely, and—maybe most importantly—they must show that transferring the authority will lead to a preferable result. The burden of proof, in other words, lies with the employees.

Nonperforming Whales, Act 2

There must be times when a killer whale just doesn't want to learn, no matter how many fish it has consumed. So our colleague asked what they did when they identified a problem whale.

"We fire it," he was told. "I mean he's probably a good whale but just not right for our whale show. So we let it go—that is, if we can't get our competition to take the whale from us. But we don't make that decision easily. It's expensive to find another whale, so we first ensure that we are not the problem. At some point, if our efforts don't succeed, you just have to cut your losses. Otherwise, it would cost $75 for the show: $30 to see the performing whales and $45 to help us feed the whales in the pool out back, where the nonperforming whales hang out. And heaven help us if the performers ever found out (and they would) that a whale could get the same treatment out back without working."

What if we shifted that burden to management? In this paradigm, the assumption would be that the people (or more likely, teams) who work the process also have complete authority over it and are held accountable for its success. Meanwhile, if anyone wants to centralize authority (i.e., move authority further up the hierarchy, assuming there is one), he or she will have to convince the organization that doing so would lead to a preferable outcome. It would also be up to the team to ask for the help it needed, submit its own budgets, measure its own performance, calculate its return on investment to the company, and generally justify that it is worthy of its members' salaries and the organization's decision to entrust it with ownership of the process.

Of course, shifting the burden of proof in this manner represents a major shift in the role of management in most organizations. While incorporating "empowered" teams into a traditional environment is a significant task, shifting the burden of proof is even more of a challenge, demanding entirely different accountability practices and an entirely different environment. Still, we feel that giving people the type of knee-knocking accountability that puts them at risk—the way that owners are at risk—is an effective alternative to a system that boasts empowerment but often delivers a great deal less.

Several years ago, we asked Ralph Stayer, then CEO of Johnsonville Foods, about team training. "Mostly unnecessary," he replied. "Just give people a real job, and the team will form and work effectively." When the responsibility is there, when the authority is "real," and when people have job competency, they will do what they have to do to deliver the value they are being held accountable for.

6. Developing Ability Commensurate with Responsibility

The fastest way to drive an employee insane is to give him or her new responsibilities and fail to provide them with necessary instruction and training to do the job.

—Ken Blanchard

Nothing is more unfair than holding employees accountable for something they can't do. Yet, in their rush to extend accountability, this is precisely what many organizations have done (and continue to do). We've asked employees, especially on the front line, to improve and be accountable for their performance. In many cases, however, we've given them little control over the process, which is frequently the cause of most of the problems that exist. In other words, the accountability is there, but the individual employee's ability to meet it is not. As a result, employees are often frustrated, customers are disgruntled, and we rarely get the improvement we're looking for.

How can this mismatch between ability and accountability be averted? How can we be certain that accountability is fair and that it supports our efforts to improve and change? Here are three questions to ask before assigning accountability:

1. *What must be done?* Before people can be held accountable, they must have a clear understanding of what it is they are being held accountable for. In any event, they shouldn't be held accountable for performing an activity. Rather, people should be accountable primarily for making a significant contribution. Also, more frequently than anyone wants to admit, people are held accountable for tasks not worth performing. Every task for which an individual is to be held accountable must be examined to ensure that it adds value. If a task is not worth doing, don't hold anyone accountable for doing it. In fact, stop doing it!
2. *Who will be accountable?* Who's responsible? Is it someone or some group? Too often, responsibility is unclear. Accountability should be assigned at the lowest level possible. Those closest to customers (or processes) are best equipped to deal with them and should, therefore, be the ones who are held accountable for the consequences of their actions. When there are teams, there should be team, not individual, accountability. Team members must win or lose as a team. We're not talking casual accountability here, but genuine "the buck-stops-here" accountability. So, when the question is asked, "Who's accountable here?" there's a single raised hand, not a half-dozen fingers pointing in every direction. This is not to say that accountability should engender fear but, rather, that each person should be responsible for his or her actions. Besides, accountability is inevitably accompanied by a certain amount of discomfort.

3. *What abilities will those being held accountable need to achieve success?* The issues raised by this question are often discussed but rarely resolved. We've seen few systematic plans to ensure that the right people in the process have sufficient control or the right skills to do the job required. This problem is sometimes compounded by a managerial mind-set that often results in managers receiving the lion's share of the benefits of training and education.

Elements of Success

What are the elements that constitute the "ability to succeed"? Clearly, there is more involved than the formal training we usually associate with making someone "able." Here are some questions to spark discussion about a person's or group's ability to perform specific tasks.

To what extent do these people or groups have the following:

Goal Clarity

- Do they understand customer and other stakeholder interests?
- Do they have a consistent vision of desired outcomes?
- Do they understand the company's strategy and the group's role in the process?

Education and Experience

- Do they understand the complexities of the task?
- Do they have the ability to anticipate potential problems?

- Do they have the ability to predict the consequences of different courses of action in a variety of circumstances?

Control of the Process

Do they have the authority to:

- Change or redesign the process as required?
- Deviate from the process when called for?
- Influence peers and other departments to achieve needed cooperation?

Information

- Do they have access to the information needed to analyze the process (understand customer needs/trends) and make decisions regarding the most effective courses of action?

Resources

- Do they have the people, financial resources, tools, space, etc., required to accomplish the task efficiently?

Time

- Do they have a reasonable time frame?

  (Note: There are no unrealistic goals, only unreasonable time frames. However, too much time pressure can limit alternatives unnecessarily, while too little time pressure can delay change efforts.)

Shared Values

- Do they have a standard for decision making?

7.
Building Cooperation Instead of Internal Competition

Are they [the people in your company] thinking more about customers or employees? About competitors in the marketplace or competitors in the hallways? About products or protocol?
Davis and Davidson, *20/20 Vision*

Teamwork. We're all for it! Many of us have taken part in expensive team-building exercises, during which we've climbed rope ladders, built boats, gone on challenging photo safaris, faced mythical enemies, and solved mysteries. Over and over, we've been told we have to learn to work together more effectively. And for good reason. Today's world is too complex for any one person to manage the value proposition alone. This is particularly the case in companies in which individuals from different functional areas, departments, and physical locations have to work in harmony to meet the unique needs of their customers. If we hope to succeed, there is, in fact, no real alternative for many of us but to work as a team, to cooperate.

Unfortunately, it is this alternative that most organizations are least well equipped to implement. We are simply not prepared to cooperate—culturally, structurally, or philosophically. We are, after all, a nation that makes heroes of individuals who distinguish themselves competitively. From corporate raiders to rock stars, from baseball's Most Valuable Player to the most famous heart surgeon, we've focused on individual winners. Looking out for number 1 is not only expected but is rewarded. Nowhere is winning (and losing) more a part of everyday life than in our organizations. Here, the thrill of victory or the agony of defeat is relived by many on an almost daily basis. Unfortunately, the primary field of battle may not be in the marketplace but in our hallways. The arena where the most intense competition in business is often waged is within the corporation itself, among its employees. The costs of unnecessary internal competition are prohibitive in most organizations.

We talk about being part of a team but, instead of cooperating in a system that encourages teamwork, we are pitted against one another to compete and win as individuals. As much as we claim we want cooperation, most of our structures don't reward it, our corporate culture doesn't support it, and our leaders are reluctant to embrace it—though it is often in the best interests of the organization to do so. Our people return from their team-building weekends and, within a few days (hours?), they're often back to building their empires at the expense of the other person and with the hope of a superior (i.e., winning) performance appraisal, higher merit pay, the next promotion, or more job security. To meet today's demands, we need to be pulling together, yet the internal competition endemic to our system is undermining our efforts.

IN SPITE OF THE SYSTEM

We're not saying that cooperation doesn't exist in most organizations. In fact, cooperation can and often does prevail. People frequently act on their more noble instincts and help one another. Indeed, it is nothing less than amazing that, given the systemic promotion of competition and the resistance to cooperation, there is as much pulling together as there is. It seems that individuals are genuinely driven to do the "right thing"—contributing to the overall good of the organization—rather than feathering their own nests. But the cards in many organizations are essentially stacked against it. There are simply too many barriers to acting decently and cooperatively and, conversely, too many rewards for acting otherwise. If our structure required teamwork, most people would readily cooperate without a library of team-building exercises. The task itself would create a cause to rally around.

The organization whose employees must compete for a limited number of wins or rewards—whether it's membership in the President's Circle or the prize for Employee of the Month—tends to create a scarcity mentality of its own, complete with the same scrambling, the same hoarding, the same subversion of potential values, and the same failure to consider the greater good. The principal difference in the type of scarcity that exists in world markets—shortages of food or medicine, for example—and that which exists in the corporation is that the former is real while the latter is artificially created. Organizational structures such as forced-distribution performance appraisal or programs such as sales and service contests get people to compete against one another—a few people win at the expense of others. Ironically, what is scarce and often bitterly fought for in the corporation is not even necessarily something that has real value in the outside world—the boss's praise, for example.

The question is, why do we do it? Why do we limit the number of people who can win? Are there great payoffs for the distinctions we make? What are the potential downside risks? Here, for starters, are eight of these risks. There are many, many more.

1. *Internal competition drives out creativity and innovation.* Individuals competing against one another must be playing the same, or similar, game so that the appropriate comparisons can be made and the winners selected. Spawned by competition, this need for similarity makes it more difficult for the individual competitor to experiment and try new methods. The result can be significantly lower levels of creativity and innovation.
2. *Internal competition inhibits dialogue.* Add the concept of winning and losing to a dialogue and you get a debate. In an internally competitive environment, individuals become less interested in sharing and thinking about new or conflicting information and become more concerned with scoring points or pressing their case, right or wrong. After all, in a competition, winning is the name of the game.
3. *Internal competition impacts relationships negatively.* Though many of us naturally expect the best from people, when we know we are competing—whether for a promotion, to curry favor with the boss, or to get a bonus—it is difficult for us to build trust, work as a team member, or create an honest relationship. This is especially so when wins are limited and our victory is made possible by the failure of others. In any event, competition makes people suspicious of one another and often results in greater anxiety in employees, who constantly feel they have to watch their backs.

4. *Internal competition lowers product and service quality.* As different individuals and different departments compete among themselves while trying to get products and services out as quickly and profitably as possible, the temptation can be to cut corners, particularly if the problems created in doing so won't come back to haunt the company for years, long after the individuals involved have moved on.
5. *Internal competition destroys focus. Winning and improvement are very different goals.* When we focus on besting others, we are not necessarily focusing on improving the present system. In fact, there is significant evidence demonstrating that winning can actually result in lower levels of performance. When people compete with a focus on winning, they often take the fastest, most reliable, most predictable route to winning, which is rarely the most effective route to continually improving the method of work.
6. *Internal competitiveness reduces efficiency.* When individuals are less innovative, less creative, less trusting, or more combative, it costs the company in a variety of ways. For example, when people compete with one another, they tend to work independently, often duplicating the efforts of others, solving problems that have already been solved, generating data for projects for which perfectly good data already exists.
7. *Internal competition demotivates the nonwinners.* The theory was that, if people competed and the winners were rewarded, they would feel appreciated and all those who didn't win would strive to do so in the future. In most cases, however, it

hasn't worked out that way. Far too many ill-thought-out and incomplete reward systems have resulted in the selection of "winners" whose performance was not any better, and was sometimes objectively worse, than that of some of the "losers." Predictably, this can be demotivating to those who "lost." Precisely how demotivating is hard to tell because we usually evaluate the motivational effectiveness (or ineffectiveness) of our efforts by tracking the reaction of the winners rather than that of the nonwinners. Instead, we simply assume that nonwinners are generally satisfied and that they'll try harder next time. In fact, the opposite often occurs. Once people are labeled nonwinners or feel that the deck is stacked against them, they sometimes stop trying altogether—and who can blame them?

8. *Internal competition lessens self-esteem.* When some people lose (or, at least, are labeled nonwinners) in the competition, they begin to question their ability to succeed in this (and maybe any other) system. After all, they worked hard, tried their best, and came up short. Losing is more common than winning because the system has been designed to make sure that most people don't win. The long-term effect of constantly coming up short can be significant. Less confident people simply don't learn or experiment as effectively as confident people.

BUILDING COOPERATION

Fortunately, there are a number of ways to break the hold that internal competition may have on our organizations:

- Increase the interactions between individuals and groups. The faceless person in another office performing another function can easily be ignored. It's far harder not to cooperate with people you know, especially if your interaction with them is frequent and you have to cooperate to get the job done.
- Ensure that everyone has the opportunity to win. This doesn't mean rewarding nonperformers. However, when a number of people seek the same goal and the number of wins is artificially limited, competition will result. If everyone can win, one person's success doesn't necessitate another's failure.
- Establish cooperation and respect as core values. As long as internal cooperation is perceived to be optional and internal competition is tolerated, little will change. Elevating cooperation to a core value will ensure that climbing someone else's back to get ahead will be career-limiting. Take a few minutes to analyze whether cooperation is career-enhancing in your organization. How do you know whether people are cooperating effectively? Do you understand the effects of the structures that promote competition? Do your human resources practices encourage teamwork or individual excellence? What happens to people in the organization who play politics at the expense of others?
- Recognize teamwork, appreciate cooperation, improve recognition abilities. The more symbolically, visibly, and frequently we demonstrate appreciation for teamwork and interdepartmental cooperation, the more of it we'll get.
- Educate everyone about the entire process. If people understand how their performance affects others, they will be less likely to act in ways that

negatively impact others or the company. Most people want to do the right thing and, given a choice, will make a good decision unless it is personally punishing. Providing information about the process enables each person to make an informed choice.

- Beware of quick rotations. A focus on short-term results often leads to a less cooperative environment. If a person has only a short time in which to impress the boss or make an impact, the long-term benefits of cooperation can be perceived to be less important. This is especially true if a person will be rotated before he or she experiences the negative effects of noncooperation. Clearly, in the fast-paced world in which we compete, most jobs don't last long and quick job rotations are necessarily the rule. In these cases, it's even more important to ensure that internal competition is limited and that accountability for cooperation is significant. If part of everyone's evaluation included an analysis of what a person or group has done to further teamwork, most organizations would be very different. Whom have you helped lately? From whom have you learned? Whom have you taught? What group has influenced your thinking the most? Which people, outside your group, have been the most instrumental in helping you succeed?
- Involve everyone in at least one cross-functional improvement effort. Mandating that people must work beyond the boundaries of their own department or functional area will result not only in more talent being applied to complicated process issues, but also in a greater appreciation of how the entire process works to benefit the customer. One of the

most important reasons for mandating participation may be symbolic, showing people that cooperation is not optional and that everyone will actively participate. People should also be held accountable for their participation. There's no greater waste of time than to be a mandated team member on a team that accomplishes very little.

- Consider teams as the primary unit of responsibility. Design in interdependence. It hardly makes sense for several members on the team to be able to win when the team as a whole loses and when the customer is shortchanged. If a task requires teamwork, ensure that teamwork is required of everyone. Make the team responsible and accountable. Beware: As some organizations have begun to organize in teams, many have been unwilling to abandon the old structure completely, keeping team members accountable as individuals to their former functional department. The result is usually a group that does not function as a team but as individuals representing an area of expertise, each with veto power over most decisions. Frustration levels tend to be high as people are torn between functional responsibilities and group commitments. If we want teams to act like teams, we must make them cohesive units, with a real task and team accountability. Going halfway can be very costly.

8.

Abolishing the Corporate Caste System

In many organizations today, a large number of employees are given the not so subtle message that they should know their place. For many, there is little in the work environment that encourages them to reinvent the way work is done or to invent new roles for themselves. In fact, often the exact opposite is communicated. People are taught that bucking the system can be career-limiting and that obedience and being perceived as a good team player are all-important. They are expected to learn their roles quickly and not to venture too far into someone else's turf.

In the traditional command and control structures where many of us began our careers, the differential treatment of different classes of employees was at the heart of maintaining order and obedience in the hierarchy. The costly by-product of these practices, however, has been reduced flexibility, less learning, and an increased feeling of helplessness among certain groups. History and common sense tell us that we are healthier and more productive when we resist the instinct (natural or otherwise) to set ourselves apart from those with whom we work.

An Unfortunate Occurrence—System

"An occurrence system? What's that," we asked—and then we wished we hadn't. It's one of those I-can't-believe-they-do-that examples of how *not* to treat people. Here are the rules of the one we encountered in 1996:

If a person is late or absent from work, he or she accumulates occurrences based on the degree of their tardiness or absenteeism. It's no-excuse system. If you're sick, you accumulate occurrences at the same rate as someone who claims to have had a flat tire on the way to work.

If you exceed a certain number of occurrences in a quarter, you are disciplined according to a standard chart. (We were told by one of the administrators of the system that treating everyone the same ensures fairness.)

Occurrence awards were given according to a published schedule:

- 1 minute to 5 minutes late—1 occurrence
- 5 minutes to 2 hours late—2 occurrences
- 2 hours late to 2 days absent—3 occurrences
- more than two days absent—4 occurrences

Of course the policies applied only to nonmanagers. In this case, that meant those front-line workers who talked on the phone to customers. Managers who did not deal directly with customers were exempt from this system and were trusted to come to work and get their jobs done.

Some employees laughed as they described their reaction to the system:

"If you are more than 5 minutes late—go to breakfast. It's the same penalty for being two hours late as it is for being 6 minutes late. I know it's absurd, but I didn't make the rules."

"If you are going to be more than 2 hours late or if your doctor's appointment runs overtime, you might as well take a day or two," etc., etc.

For many others, however, the system was the source of pain, frustration, and anger:

"I was having lunch with my supervisor. It was time well spent in that we discussed good ideas about how to improve the way we do our jobs. However, we returned to the office 3 minutes past the end of my lunch break, and that same supervisor came to my desk 10 minutes later with an occurrence slip. Just when you think you can trust them..."

"I was seriously ill and they sent me a warning letter about the number of occurrences I was accumulating. I told them that I was sick—not taking a paid vacation. They told me that it was my problem. They said they hoped I got well, but that they want to ensure equal treatment for all. Is it really equal treatment to penalize a sick employee as though he just wants to take a few days off?"

Who thinks this stuff up? We ranked this system right up there with such past telecommunication practices as having operators sign out to go to the bathroom, a practice that was still in effect in some companies until quite recently. Although systems like this one are rare, the mind-set that created them is alive and well in many service organizations.

Shortly after we discovered the occurrence system, the management of the organization changed. The group that took control was given a mandate to improve employee retention and to build a system that would help people make a maximum contribution to the business. When we told the new management about the occurrence system, their first reaction was to put a team together to study the potential effects of disbanding the system.

Wrong Answer!

Providing inspiration and impetus for the past successes of many of our leaders has often been the distinctly American notion that anyone can be anything he—and, in recent years, she—wants to be, that anyone can grow up to be president. Where you were born is regarded as far less of a factor in your success in this country than in others. After all, America was built by men and women who transcended their past and who believed they could change their fate. Refusing to know their place, those who came before us dreamed a dream of unlimited possibilities and created an unparalleled level of innovation.

Our recent entrepreneurial history suggests that this dream is still alive for many. However, for a growing number of people locked in certain jobs (particularly in the service sector), the American dream is just that—a dream. Where you went to school, if you went to school, where you entered the organization, whom you worked for, and which department trained you can brand you and greatly influence your ability to grow and succeed.

The unfortunate fact is that the first impression you give may make or break your career in many companies. We've all either been there or seen it happen: The boss takes a liking to us and, therefore, gives us more responsibility, treats us as an important team member, delegates decision-making responsibility, and works hard to ensure that we are supported. The not so subtle communication is that we've been selected as a top performer and, not surprisingly, we often succeed. Even if we fail along the way, the person who "marked us for success" has a vested interest in ensuring that his or her judgment was correct and bails us out.

On the other hand, we have all seen what can happen when people are branded with less favorable endorsements. They get less interesting assignments, less responsibility, less decision-making authority, and less positive support. For some employees, the result seems inevitable. They are treated as part of the underclass, and success is substantially more difficult for them to achieve.

While some of the messages about our status are subtle, many are often explicit and unambiguous. Try eating in the executive dining room if you don't rate. Have we ever figured out why many promotions necessitate a slightly larger office, fancier desk, better furnishings, and maybe even "real" art? And just listen to what we call one another—"boss," "subordinate," "manager," "bargaining unit," or "hourly employee." The meaning underlying our words is worth pondering. Is it indicative of a mind-set that demands visible differences in status to maintain control? We once worked with an executive who was willing to spend millions of dollars to reorganize the entire company to, as he put it, "break down barriers to communication." But he was absolutely unwilling to discuss giving up his assigned parking place. In too many organizations, having the indices of success is perceived to be more important than the quality of one's performance.

Our roles in perpetuating status differences in our organizations may be difficult to evaluate until we experience what it is like to work without them. It is hard to explain the learning that can accrue when you enter a world where status does not play as large a role. Touring Honda's assembly plant in Marysville, Ohio, amid a sea of white lab coats one day, we were totally immersed in a conversation with a man we assumed was a front-line team member. An hour later, our host introduced us to this individual as a Honda executive, leaving us to consider all the reasons why we had thought he worked on the front line rather than in management and how our new knowledge altered the way we now perceived him. In fact, we had always prided ourselves on not being significantly influenced by a person's position.

But we left Marysville with uncomfortable questions about our assumptions. The full effect of the subtle (and often not so subtle) messages created as a by-product of treating people differently within an organization is sometimes not readily apparent to those who work there. If, however, we can step back far enough to get an objective view, it becomes quickly evident that the unintended caste system that has evolved undermines our ability to capture the potential of the workforce. Class distinctions tend to create bureaucracy, lessen self-esteem, and lead to a feeling in many individuals that they have little control over the systems in which they work.

Downsizing and flattening organizational hierarchies might help cut down on the negative effects of class distinctions but, if companies don't change the structures that created the problems initially, it won't be long before a similar caste system emerges. But why? What will it buy us? The benefits of these practices are hard to comprehend, and yet there is little doubt that they can rob the organization of the very values it must have to prosper in the future. We need to create structures that communicate that people should not know their place. We need people eager to remake their roles in search of new ways to make a meaningful contribution.

The good news is that, in organizations where things have to get done, and done quickly, people have less time for, and less patience with, the trappings of status. In these organizations, the demonstrated ability to contribute is everything, and class and status are next to nothing. The challenge then is for each of us to choose to make it increasingly more difficult to differentiate unfairly or unnecessarily among our fellow employees.

9.

Developing an Optimistic, Caring, and Supportive Environment

Life inflicts the same setbacks and tragedies on the optimist as on the pessimist, but the optimist weathers them better. As we have seen, the optimist bounces back from defeat, and, with his life somewhat poorer, he picks up and starts again. The pessimist gives up and falls into depression. Because of his resilience, the optimist achieves more at work, at school, and on the playing field. The optimist has better physical health and may even live longer. Americans want optimists to lead them. Even when things go well for the pessimist, he is haunted by forebodings of catastrophe.

—Martin Seligman, *Learned Optimism*

Most of us believe that optimism in the workplace can be a blessing, pessimism can be a problem, and helplessness—that is, the sense that there's nothing we can do to improve our situation—is a feeling to be avoided. Many of us even know an optimist or two—a persistent salesperson who won't take no for an answer and is undeterred by the inevitable string of rejections that comes with the job or the successful, seemingly obsessed leader who persists in doing what everyone tells him or her is impossible. Still, there is strong evidence that more and more people are beginning to feel pessimistic about their ability to impact the organization.

What Do You See When You Look in the Mirror?

Just as people will react to their perception of the leader's expectations of their abilities, they will also react to their perception of the leader's self-confidence. If people do not believe that the leader is confident in his or her abilities to lead the organization in the espoused direction, people will be hesitant to follow.

Caution: People have good BS detectors. They are not easily fooled. When you look in the mirror, do you see someone who is passionate, confident, and optimistic about where the organization is headed?

Although we still believe in the importance of optimism, and although it is critical to our efforts to adapt in continuously changing times, it is becoming an increasingly rare commodity in some organizations.

The more uncertainty there is, the more complex the issues we face, and the more experiments we must conduct, the greater the number of mistakes that will inevitably result. To the optimist, these mistakes are speed bumps, necessary minor irritations, challenges on a road that ultimately leads to a desirable outcome. Optimists tend to be more persistent and less affected by unanticipated difficulties. Their solutions tend to be more imaginative and creative, and they are usually more avid learners.

Caring is Not Optional

Do reward and recognition events work? Do they motivate? Do they really focus efforts on the attainment of specific corporate goals? Do they communicate to employees genuine caring on the part of management? More often than not, our experiences tell us, they don't.

More often than not, employees are all too aware of the strings attached, of the underlying manipulation, and of having to play out a scene in which the lines are forced and insincere.

Sometimes, if recognition events are taken seriously, they do positive harm. By rewarding a handful of "winners," management leaves the majority of employees feeling like losers. Often, too, the "losers" resent management's choices as based on politics more than merit, and the awards become tangible symbols of favoritism, flawed measurement systems, and bossism.

Recently, while working with a company to assess the effectiveness of their reward and recognition practices, we were told firsthand of the problems with what a number of employee groups referred to as "Trinkets and Trash" programs. The mere mention of recognition events elicited comments that might generously be described as cynical.

The gap between management's "walk" and "talk" was fuel for a seemingly endless list of stories that gave evidence of management's dwindling credibility. Listening to these employees, it was obvious that leadership inconsistencies lay at the root of their cynicism. Not that these employees turned down the money or plaques or the free dinners they were awarded: They accepted them all, but not in the spirit that management had hoped for. We had to wonder what the long-term effect of these events would be. We asked these same individuals why they thought management would go to the trouble of investing in these events if they didn't care. "They're doing it to fulfill the Baldrige criteria," was a common response, as was, "They think this stuff motivates us." Some employees attributed the best of motives to individual managers, but hypothesized that the way the business was organized made it very hard for managers to care about people and still get ahead.

Others believed that many managers viewed employees as a resource that had to be "managed" on the way to the really important stuff—shareholder return and stock price. When we asked managers in this company why they held these events, they were generally consistent in their stated motives. "Appreciation for past performance" and "the events' motivational effects in guiding future performance" were at the top of most lists. By and large, we found these managers to be sincere, caring, and well-intentioned. However, their failure to walk their talk on a day-to-day basis, a list of organizational structures that were perceived to be inconsistent with stated goals, the suspicion among employees that management had an ulterior motive, and a perceived lack of genuine caring by management turned many of these acts of appreciation into opportunities for building peer alliances and bashing senior management.

Compare this company's recognition events to those we witnessed at Disney. Outwardly, the events at both organizations were similar. The speeches echoed many of the same themes, and the activities and meals were virtually indistinguishable. But the reactions of those in attendance were as different as night and day. In the first company, employees were cynical about the events and distrustful of the 'sponsors' motives. In contrast, most Disney cast members viewed their events as genuine celebrations, thank-you's from the company's leaders who, the cast members felt, truly cared about them.

We're not suggesting that the celebrations at Disney were perfect or that Disney has no disgruntled cast members. But the overall mood and atmosphere at the Disney celebrations were markedly different from those at the other company. At Disney, the number of celebraters was surprising, there was a far greater number of "winners," and the involvement of individuals from all levels of the workforce was generally the rule, not the exception. Celebrations seemed to be part of the day-to-day culture of the organization and were, therefore, less likely to be perceived as an infrequent event held to "get something."

We can't pretend to know whether these employees' interpretations of management's motives were correct in every instance. We've found, however, that most people are pretty good at detecting motives. If we are ever tempted to fool those we work with, we'd be well advised to think again. That's why, the next time we plan a recognition event, we should first ask:

- Do we personally care about these employees? Or are we just interested in what they can do for us?
- Do we routinely support their efforts to improve, learn, and grow, or are these goals secondary to what we want from them?
- Are the structures of the organization consistent with the values we will espouse in this event? If not, have we visibly committed to changing them?
- Have we made a firm commitment to telling the truth? Considering most organizational histories, people have a right to be skeptical. If we fudge on the truth even a little, we're in deep trouble.

As we've said before, every organization is inescapably human. Caring is not optional. Caring, not manipulation, is the currency most likely to build a good foundation for the kind of team we will need to succeed in the future.

People don't care how much you know until they know how much you care.

John Hanley, President and CEO, Lifespring

On the other hand, to the pessimist, or to someone who has learned to feel helpless, mistakes don't seem like speed bumps. To these individuals, mistakes can appear as the unexpected end of the road. Demoralized, they can't even imagine finding an alternate route to success. For some, even the smallest problem is felt to be a disabling event over which they have little, if any, control. Predictably, these individuals expend less effort on the job and their successes are infrequent.

If there were only a few pessimistic employees in our organizations, their effect might not be so crippling. The fact that there are many—it is said that two-thirds of Americans today tend to be pessimists—means that the performance of virtually every corporation has been impacted. Whether the feelings of helplessness that often accompany pessimism have been learned outside the workplace or on the job as a result of past practices, employee pessimism is a gargantuan value subtracter. This is particularly the case as we increasingly rely on employees at every level to learn, experiment, and use data to innovate and improve performance. We must, therefore, refuse to give in to pessimism. We must learn to be optimistic and to help others do the same. We must also redesign those organizational structures that create a work environment in which individuals have little control and where a feeling of helplessness is perpetuated.

Unfortunately, many traditional management practices contribute to creating and perpetuating precisely such an environment. For many employees, especially those at the base of the organizational hierarchy, narrow, rigid job descriptions communicate that the individual can have little effect on the design of his or her work. Reinforcing this message are job demands for which workers receive insufficient training, information, and resources. When employees perceive that they are not trusted or that their past efforts have not been valued, their tendency to be optimistic is inhibited and their pessimistic tendency is encouraged. Traditional practices that lead to finding fault with people instead of processes, dismantling improvement efforts before they can be successful, limiting feedback, perpetuating inflexible procedures, and creating a corporate caste system that tells employees to know their place are counterproductive. The cumulative effect of these practices fuels a sense of helplessness in many employees and reduces their ability to persevere in the face of the resistance inherent in most change efforts.

If change is what we are after, we have little choice, therefore, but to eliminate those structures that tend to increase feelings of helplessness in people. We need to build employee confidence and provide individuals with a sense of control over their environment. People should have the opportunity to participate meaningfully throughout the organization and have a significant say in the way work is done. Substantial involvement is not optional but is a prerequisite to building an effective, optimistic team. Optimistic people and teams learn more, achieve more, and are more successful. Optimists usually lead happier and healthier lives and have a lot more fun.

Nevertheless, optimism isn't always preferable. Because pessimists tend to see the world more as it is, and optimists perceive it more as it suits their needs, being an occasional pessimist can be a real benefit. If, for example, you're considering a decision that "bets the company," it's better to be somewhere along the continuum between realism and pessimism. However, in approaching most tasks, the advantages of optimism can be substantial and, in many organizations, are largely unrecognized. We're not suggesting that it will be easy to make optimists out of people who have learned to be more pessimistic—or turn yourself into an optimist if you're not. But, because optimism and pessimism are states of mind, they can be influenced and ultimately changed. All of us must resist the temptation to give in to the feeling that our actions don't matter.

TURNING FAILURES INTO LEARNING OPPORTUNITIES

To characterize a mistake as a learning opportunity—as has often been recommended in recent years—does little good unless the person who is doing the characterizing believes it to be true. As a general rule, optimists will and pessimists won't. The more optimistic a person is in his or her ability to solve a problem, the greater the chance that that person will view failure as an opportunity. However, people who believe that the problem and its causes are out of their control, that they are helpless to affect the result, and that they might even be at fault are less able to learn from their mistakes and take corrective action. How a person perceives a given situation is everything. Putting a different label on a set of facts doesn't automatically make things better. If we want to encourage people to see mistakes or failures as opportunities, we must create an environment that supports this concept and that communicates to all employees that:

- Failure is temporary. The result of any failed attempt can be overcome with time and hard work.
- Failure is natural. In most complex endeavors, there will be more failures than successes. Succeeding only once out of 10 tries might be a perfectly acceptable average in some situations.
- Failure is probably not their fault. Yes, we should take responsibility for our actions. But a majority of the problems we encounter on the job are precipitated by the way work has been organized, not through the fault of individual participants.

- Failure is the key to learning. Every unsuccessful attempt is a chance to increase our understanding of the system and to experiment further. Failing doesn't mean giving up; it means looking for alternatives.

Martin Seligman, author of *Learned Optimism,* has said that each of us carries a word in our heart. For some of us, that word is *yes.* Yes, we believe we can succeed. Yes, we can learn. Yes, we can make a difference. Others carry a *no,* with all the negative baggage that accompanies it. As leaders, we must realize which word we carry and how it enhances or inhibits our ability to lead. Skills and desire are not enough to succeed. We must also be convinced that we can...and be able to convince others, as well. If we are passionately optimistic, odds are that our optimism will be contagious. In any event, our improvement efforts, no matter how well considered, will have little chance of succeeding if a significant part of the workforce feels powerless to make a difference.

10.
Building Trust One Employee at a Time

...You cannot survive in a global economy with low quality. And you cannot produce high quality unless you have high trust. So it is also impossible to have high, sustainable trust unless you have deep trustworthiness in the people of the organization.

—Stephen Covey

For years, we've talked the tough talk about our need to create a fast, flexible, adaptable organization. To be sure, we've taken a number of positive steps in that direction. Still, there are many barriers to realizing that goal. Perhaps the single largest of these is also the least visible and one we're least likely to question: our own lingering mass-mentality mind-set that personalization in the workplace is simply too expensive and that we can reach our objectives without having to deal with complex and frequently messy personal interactions and relationships. Compounding the problem posed by this barrier to a more productive workplace is our own seeming inability to do anything about it, even when our best instincts tell us that One Size Fits One relationships are the *only* way to build commitment—and that commitment has to be built individually, one employee at a time.

The Signs of Effective Leadership are Found in the Followers

If our jobs as leaders are to help create a passionate, committed team focused on creating value for customers, than why wouldn't we judge our successes or failures by evaluating the extent to which the team is focused and performing? Too often, leadership effectiveness has been judged by evaluating the leader's behavior without enough attention paid to whether that behavior worked with the team. All leaders have at least one thing in common—they have committed followers. Without them, whatever the leader did was ineffective.

Then, again, being unable to act on our own best instincts in an organizational setting is hardly surprising. It's always been difficult for *anyone* to foment real change in the organization or take action based on a mind-set that challenges established practices as One Size Fits One employee practices most certainly will. After all, virtually everyone in the organization has been taught to think in terms of the mass, not the individual. We've all been given objective (i.e., nonmessy, non-relationship-oriented) goals to meet, most of them short-term. In realizing these goals, our reward systems have always encouraged us to do more of what we did yesterday. Our mass-mentality-derived structure and the practices it has spawned have taught us the types of behavior that can enhance our career and the types that can limit it. Besides, in most cases, the system we live by is precisely the system that has anointed us with the success we enjoy. And, while we may have major misgivings about that system—or are beginning to question its utility and worth—at least we understand it and know how to work it in our favor.

The Role of Trust at Sun Microsystems

I trust our people, period. We don't have a process book or procedure for everything. We have trust.

—Mel Friedman, Vice President
Worldwide Operations, Sun Microsystems

In fact, trust is so important an element at Sun Microsystems that it appears as a measurement on the performance appraisal system. Friedman told us that he also struggled with the concept, because, in his mind "You either are trusted or you're not. You either choose to trust employees or you don't." Friedman said he trusted people at Sun because he believed that they were all trying to do the right thing. Other members of his worldwide operations team, in the room during our interview, voiced similar beliefs. They said that, when issues crop up, they always know and trust that there are no hidden agendas, just different approaches to issues. The basic trust at work is that everyone on the team, at all levels, and throughout the organization, is trying to do the right thing.

Friedman gave us an example of how ingrained the *lack of trust* in people often is in the high-technology industry. It is well-known that, in the valley, when an employee gives notice of resignation, he or she is promptly escorted from the building. We suppose this "tradition" evolved because of the trade secrets and technology information inherent in the industry. A manager who reported to Friedman resigned to take another position on the East Coast, where his wife had been transferred. After announcing his resignation, the employee said to Friedman, "Well, aren't you going to escort me out of the building?" Friedman replied, "You're no different a person to me today than you were yesterday. Why should I escort you out of the building? I trust you. Besides, if you were going to take any secrets, I also trust you would have been smart enough to have taken them before now."

So, to say that it's going to be easy to break away from this familiar way of going about our business would be to turn our backs on the system that we grew up with and that rewarded us on the way—a system we both know and, maybe, in some weird way, even feel comfortable with. Still, we can't pretend that a system founded on people as machine parts is either particularly human or humane. We can't, in any event, pretend that it maximizes human creativity and productivity. We can't because we know the system all too well and know for a fact that this is simply not the case.

The question is, then, given how deeply entrenched the mass mentality is in most organizations, do we even have a chance to make the changes we have to make to win the commitment of those we have to have on our side—customers and employees alike—as we face the future? Even if we want to change, *can* we?

EARNING TRUST TO REALIZE CHANGE

There is no doubt about it, particularly if the need presents itself in an urgent and undeniable manner—as in a diminishing customer base or an eroding bottom line. But, once we ourselves are genuinely committed to tapping into the power and energy of the workforce by creating a more human organization, to be successful in our efforts for change, we have to be prepared to encounter yet another obstacle...and it won't be the recalcitrance of our systems. Nor will it be the organization's knee-jerk resistance to change of any sort. Beyond our own mind-set, our primary stumbling block to real change will be the far more human, far less predictable, and more difficult to quantify element of *trust*—the trust of our co-workers whose commitment we seek, as well as our reciprocal trust of them.

Let Go

When we asked the CEO of a billion dollar furniture manufacturer what advice he could give managers hoping to unleash the power of the workforce, his answer was short and to the point:

"LET GO! You must let go.

"It will never be easy, but you must do it anyway. When I took over, I was lucky. I had no choice. Of course, I felt out of control when I no longer made all the decisions. I didn't know for sure what would happen—how many mistakes would be made. Of course, I felt like I could have made some of the decisions more effectively, but I learned that, if you give people a chance to learn to experiment, to alter the course of the business, they can take you to heights that you might never have imagined possible.

"But first, you have to trust that it can and will happen. Then you can let go."

Without that mutual trust, any attempt on our part to set a new course will inevitably be frustrated and the mass mentality born of the Industrial Revolution—and its mass-production practices—will continue to have more control over us than we will over it. With trust, however, change is not only possible but likely, as individuals unite in trusting partnership with us in pursuit of a mutually compelling goal.

As we all know, however, trust is not easy to come by. You're either trusted or you're not. You either feel it or you don't. Trust takes years to win and minutes to lose—and chances are, over the last few years, we've probably lost a whole lot more trust with our co-workers than we've gained by telling them—or our customers—we cared about them as individuals and then bulk-handled them through one-size-fits-all employee systems. This is why our first order of business on the road to change will probably be to win back that trust, meaning:

- To say what we mean and do what we say.
- To live by our values.
- To give and take feedback and criticism without rancor or hidden agendas.
- To forsake company politics.
- To make ourselves vulnerable—even in the face of leadership principles that tell us to keep our distance so that we can't be taken advantage of.
- To care genuinely for those we work with and those we serve and to freely demonstrate our caring through our actions.

Now, if this sounds less like "business as usual" and more like "touchy-feely," it's only because it is. There's no doubt about it, touchy-feely and the emotions that surround it are precisely what forming human relationships is all about.

"SHALL WE GET ON WITH THE JOB?"

In the end, if we're ever going to realize huge improvements in our organization by leveraging the talents and commitment of each individual on the workforce to build those relationships with our customers that we so desperately need to remain competitive, then we have little choice but to immerse ourselves in the essential "humanness" of our organization. Doing so, however, will require enormous changes in the way we do business and in the way we lead. Still, if we don't take on the challenge, who will? And if we don't start immediately, will we ever?

INDEX

The Center For Innovative Leadership

- The Center For Innovative Leadership, is a consortium of educators, consultants, and communications professionals dedicated to helping leaders unleash the potential of the workforce. The Center achieves this objective through research, writing, training, and consulting.

We would like to hear from you! Please contact us in the following ways:

The Center For Innovative Leadership
1400 Fashion Island Blvd., Suite 601
San Mateo, California USA 94002
1-800-878-3601

or through our World Wide Web page and email addresses:

WORLD WIDE WEB PAGE: http://www.CFIL.com

EMAIL TO THE AUTHORS:

Gary Heil @dsp.com
Deborah S @dsp.com
ANNAANN@aol.com (for Tom Parker)

We are currently gathering information, research and stories in the following areas and encourage you to contact us:

- Service Stories: Give us your best and your worst service story or encounter.
- New Ways Of Working: How you are coping with the "new employment contract" where lifetime employment is no longer a guarantee. How are you taking career matters into your own hands.
- Journal and Newsletter Articles: Great ones you've read and would recommend. Ones you've written and would like us to consider for publication.